United States Nuclear Regulatory Commission

Protecting People and the Environment

NUREG-2152

Computational Fluid Dynamics Best Practice Guidelines for Dry Cask Applications

Final Report

Office of Nuclear Regulatory Research
Office of Nuclear Material Safety and Safeguards

AVAILABILITY OF REFERENCE MATERIALS
IN NRC PUBLICATIONS

NRC Reference Material

As of November 1999, you may electronically access NUREG-series publications and other NRC records at NRC's Public Electronic Reading Room at http://www.nrc.gov/reading-rm.html. Publicly released records include, to name a few, NUREG-series publications; *Federal Register* notices; applicant, licensee, and vendor documents and correspondence; NRC correspondence and internal memoranda; bulletins and information notices; inspection and investigative reports; licensee event reports; and Commission papers and their attachments.

NRC publications in the NUREG series, NRC regulations, and Title 10, "Energy," in the *Code of Federal Regulations* may also be purchased from one of these two sources.
1. The Superintendent of Documents
   U.S. Government Printing Office
   Mail Stop SSOP
   Washington, DC 20402–0001
   Internet: bookstore.gpo.gov
   Telephone: 202-512-1800
   Fax: 202-512-2250
2. The National Technical Information Service
   Springfield, VA 22161–0002
   www.ntis.gov
   1–800–553–6847 or, locally, 703–605–6000

A single copy of each NRC draft report for comment is available free, to the extent of supply, upon written request as follows:
Address: U.S. Nuclear Regulatory Commission
         Office of Administration
         Publications Branch
         Washington, DC 20555-0001
E-mail: DISTRIBUTION.RESOURCE@NRC.GOV
Facsimile: 301–415–2289

Some publications in the NUREG series that are posted at NRC's Web site address http://www.nrc.gov/reading-rm/doc-collections/nuregs are updated periodically and may differ from the last printed version. Although references to material found on a Web site bear the date the material was accessed, the material available on the date cited may subsequently be removed from the site.

Non-NRC Reference Material

Documents available from public and special technical libraries include all open literature items, such as books, journal articles, transactions, *Federal Register* notices, Federal and State legislation, and congressional reports. Such documents as theses, dissertations, foreign reports and translations, and non-NRC conference proceedings may be purchased from their sponsoring organization.

Copies of industry codes and standards used in a substantive manner in the NRC regulatory process are maintained at—
   The NRC Technical Library
   Two White Flint North
   11545 Rockville Pike
   Rockville, MD 20852–2738

These standards are available in the library for reference use by the public. Codes and standards are usually copyrighted and may be purchased from the originating organization or, if they are American National Standards, from—
   American National Standards Institute
   11 West 42nd Street
   New York, NY 10036–8002
   www.ansi.org
   212–642–4900

United States Nuclear Regulatory Commission

Protecting People and the Environment

NUREG-2152

Computational Fluid Dynamics Best Practice Guidelines for Dry Cask Applications

Final Report

Manuscript Completed: September 2012
Date Published: March 2013

Prepared by:
Ghani Zigh and Jorge Solis

Office of Nuclear Regulatory Research
Office of Nuclear Material Safety and Safeguards

ABSTRACT

Dry storage cask designs for spent nuclear fuel are submitted to the U.S. Nuclear Regulatory Commission (NRC) for certification under Title 10 of the *Code of Federal Regulations* (10 CFR) Part 72, "Licensing Requirements for the Independent Storage of Spent Nuclear Fuel, High-Level Radioactive Waste, and Reactor-Related Greater Than Class C Waste." The NRC staff technical review of these designs is performed in accordance with 10 CFR Part 72 and the "Standard Review Plan (SRP) for Spent Fuel Dry Storage Systems at a General License Facility" (NUREG-1536, 2010). To ensure that the cask and fuel material temperatures of the dry cask storage system will remain within the allowable limits or criteria for normal, off-normal, and accident conditions, a thermal review is performed as part of the application's technical review. Recent applications increasingly have used thermal-hydraulic analyses and computational fluid dynamics (CFD) codes (e.g., FLUENT) to demonstrate the adequacy of the thermal design. Therefore, in cooperation with the Division of Spent Fuel Storage and Transportation of the Office of Nuclear Material Safety and Safeguards, the Office of Nuclear Regulatory Research developed this guide to provide practical advice for reviewing CFD methods used in vendor applications and for achieving high-quality CFD simulations of a dry cask. To assist in the analysis, the report includes procedures, analysis methods, and acceptable assumptions.

Deficiencies and inaccuracies of CFD simulations can be related to a wide variety of errors and uncertainties. An error is a recognizable deficiency that is not caused by a lack of knowledge; an uncertainty is a potential deficiency that is caused by lack of knowledge. An error is something that can be removed with proper care, effort, and resources; an uncertainty cannot be removed because it is rooted in a lack of knowledge. This report addresses two categories of uncertainties and provides specific guidelines to minimize them. The first category is modeling uncertainties. The difference between the real flow and the exact solution of the model equations creates these uncertainties. This report provides validation of the modeling approaches used to represent the heat transfer and fluid flow in a dry cask to reduce modeling uncertainties. In particular, the discussion of modeling uncertainties focuses on turbulence modeling, which could greatly influence the predicted results if not applied correctly. Commercial CFD codes have many turbulence models that are not generalized; therefore, they cannot be applied to all types of flows. Depending on the flow characteristics, some models are more applicable than others. Several approaches to model air flow turbulence have been investigated and compared to experimental data, and the results are included in this report. The report provides CFD best practice guidelines (BPGs) to minimize turbulence modeling uncertainties for dry cask analysis.

The second category of uncertainty relates to application uncertainties. These uncertainties are introduced because the application is complex and the precise data needed for simulation are not always available. In this report, the discussion of application uncertainties focuses on the inlet and outlet boundary conditions of ventilated dry storage casks. Usually, boundary conditions and assumptions used by applicants to perform CFD analyses are verified and compared with other equivalent approaches to reduce application uncertainties. As cooling air is naturally induced in ventilated dry casks (VDCs), specifying pressure at the boundaries is the preferred choice at the inlet and outlet ducts. The pressure gradient in the air flow channel affects the magnitude of the potential buoyancy forces because of the heat source (i.e., spent fuel decay heat). As such, the pressure boundary conditions are crucial to the uncertainties that could be introduced in the simulation. The effect of the pressure boundary condition was investigated and compared to experimental data to minimize application uncertainties. The report provides specific guidelines to avoid application uncertainties that could arise in the specification of the pressure boundary conditions at the inlet and outlet ducts of a VDC.

Recently, there has been a growing awareness that computational methods can prove difficult to apply reliably (i.e., with a known level of accuracy). This is, in part, because CFD is a knowledge based activity and, despite the availability of the computational software, the knowledge base embodied in the expert user is not available. This has led to a number of initiatives that have sought to structure existing knowledge in the form of best practice advice. Four notable examples are the BPGs developed by the European Research Community on Flow, Turbulence, and Combustion (ERCOFTAC); the European Thematic Network for Quality and Trust in the Industrial Application of CFD (QNET-CFD); the Organisation for Economic Co-operation and Development /Nuclear Energy Agency/Committee on the Safety of Nuclear Installations CFD working groups; and the American Society of Mechanical Engineers (ASME), "Standard for Verification and Validation in Computational Fluid Dynamics and Heat Transfer" (ASME, 2009). The QNET-CFD knowledge base is currently under the control of ERCOFTAC. The guidelines presented here build on the work of these four initiatives, particularly the ERCOFTAC BPGs that have been used as a template for these guidelines (with some modifications and adaptation).

This document considers the use of CFD programs solving the Reynolds-Averaged Navier-Stokes equations on both structured and unstructured meshes, as well as the use of large eddy simulation and detached eddy simulation. The report attempts to cover the full range of issues associated with a high-quality CFD analysis. It begins with a proper definition of the problem to be solved, thus permitting selection of an appropriate simulation tool. For the probable range of tools, the report provides generic guidance on selecting physical models and on numerical issues, including the creation of an appropriate spatial grid. To complete the analysis, the report also provides guidance on how to verify the input model, validate results, and document the process.

CONTENTS

LIST OF FIGURES

LIST OF TABLES

ACRONYMS AND ABBREVIATIONS

BPGs   best practice guidelines
BWR   boiling-water reactor
CFD   computational fluid dynamics
CPU   central processing unit
CSNI   Committee on the Safety of Nuclear Installations
DES   detached eddy simulation
DNS   direct numerical simulation
DO   discrete ordinate
DTRM   Discrete Transfer Radiation Model
ECORA   Evaluation of Computational Fluid Dynamic Methods for Reactor Safety Analysis
ERCOFTAC   European Research Community on Flow, Turbulence, and Combustion
GAMA   Working Group on the Analysis and Management of Accidents
Gr   Grashof
IAEA   International Atomic Energy Agency
kW   kilowatt
LES   large eddy simulation
MSB   multi-assembly sealed basket
NEA   Nuclear Energy Agency
NRS   Nuclear Reactor Safety
OECD   Organisation for Economic Co-operation and Development
PCT   peak cladding temperature
PDE   partial differential equation
PIRT   phenomena identification ranking table
PWR   pressurized-water reactor
RANS   Reynolds-Averaged Navier-Stokes
RSM   Reynolds stress models
QA   quality assurance
QNET-CFD   A Thematic Network for Quality and Trust in Industrial Applications of CFD
SIMPLE   semi-implicit method for pressure-linked equations
SST   shear stress transport
TCs   thermocouples
TRANS   Transient RANS
URANS   unsteady Reynolds-Averaged Navier-Stokes
V&V   verification and validation
VCC   ventilated concrete cask
VSC   ventilated storage cask

1.0 INTRODUCTION

Computational fluid dynamics (CFD) has developed over the last 25 years into a reliable tool for analyzing complex flow situations. For example, CFD has become an invaluable aid in design practices in the automotive, aerospace, and turbo-machinery industries. However, CFD is not as mature a technology as what is available in commercial codes for thermal and stress analysis in solid structures. The main difficulty is that industrial-type CFD is highly nonlinear and requires resolution of flow structures spanning a wide range of scales (e.g., boundary and free-shear layers, vertical structures, and zones of recirculation). Although universities and Government laboratories may continue to pursue in-house CFD development, this activity is strictly limited to departmental specialties. From an industrial standpoint, commercial vendors of CFD software now are undertaking major steps forward in CFD technology. The major players in this league are CFX, FLUENT (both now owned by ANSYS), and STAR-CD. Worldwide, current estimates of regular users of commercial CFD codes are 25,000 to 30,000, and the number has been growing steadily by 15 percent to 20 percent annually for several years. This growth has enabled the major CFD vendors to sponsor and, more generally, to become actively involved in the development of innovative numerical modeling techniques that they hope will convert into profit-based growth in the future. Examples of these efforts include direct funding of master and doctoral programs at universities and direct participation in research programs funded by the European Union.

The availability of robust commercial CFD software and high-speed computing has led to increased use of CFD as a solution for fluid engineering problems across all industrial sectors. The nuclear industry is no exception.

Recently, there has been a growing awareness that computational methods can prove difficult to apply reliably (i.e., with a known level of accuracy). This is partly because CFD is a knowledge based activity and, despite the availability of the computational software, the knowledge-based embodied in the expert user is not available. This has led to many initiatives that have sought to structure existing knowledge in the form of best practice advice. Four notable examples are the best practice guidelines (BPGs) developed by the European Research Community on Flow, Turbulence, and Combustion (ERCOFTAC), the European Thematic Network for Quality and Trust in the Industrial Application of CFD (QNET-CFD), the Organisation for Economic Co-operation and Development/Nuclear Energy Agency/Committee on the Safety of Nuclear Installations CFD working groups, and the ASME V&V 20 standards (ASME, 2009). The expertise and knowledge-based that originated in QNET-CFD was handed over to ERCOFTAC control. The guidelines presented here build on the work of these four initiatives, particularly the ERCOFTAC BPGs, which have been used as a template for developing these guidelines (with some modification and adaptation).

The range of CFD tools available for dry cask applications is broad and varied. This has presented somewhat of an issue in developing these guidelines. However, there are many common elements regardless of the tools being used. Common challenges that the CFD user in the maritime sector faces with his or her counterparts in other engineering fields include the need to understand the physics of the problem at hand, the limitations of the equations being used, the basis of the numerical methods employed, and the means to get the most accurate and consistent results for the available computing resource.

The general picture emerging is that CFD is rapidly expanding, with a large database of proven capability. The driving force for program development generally is not the nuclear community, as it was for the classical thermal-hydraulic system codes. Nonetheless, many application areas overlap with those associated with dry cask applications: flows in complex geometries, mixing in stratified fluids, flow separation and re-attachment, turbulence, multiphase phenomena, chemical species interaction, and combustion. Consequently, practitioners in areas related to dry casks applications can benefit indirectly from advancements in the technology taking place elsewhere. However, because of the complexity of modern commercial CFD packages, careful input preparation and solution of model equations are essential to avoiding errors. Some of these points are expanded upon in this document.

This document provides best practice guidelines for undertaking simulations used to evaluate the thermal response of dry casks. Dry cask applications include transfer, transport, and storage. First, the different sources of errors and uncertainties known to occur in numerical simulation results are listed and defined. The sources of error that can be controlled and quantified by the user are then discussed in detail, and BPGs for their reduction and quantification are given. These BPGs are based on available guidelines as much as possible.

For evaluating CFD codes, all of the errors and uncertainties that cause the results of a simulation to deviate from the true or exact values must be identified and treated separately, if possible. Several classifications of these well-known errors and uncertainties exist. The most general discrimination divides them into the following two broad categories (Coleman, 1997):

- errors and uncertainties in modeling the physics
- numerical errors and uncertainties

The errors and uncertainties in modeling the physics arise from the assumptions and approximations made in the mathematical description of the physical process:

- simplification of physical complexity
- usage of previous experimental data
- geometric boundary conditions
- physical boundary conditions
- initialization

Numerical errors and uncertainties result from the numerical solution of the mathematical model. The sources for the numerical errors and uncertainties are:

- computer programming
- computer round-off
- spatial discretization
- temporal discretization
- iterative convergence

When performing validation simulations, it is mandatory to quantify and reduce the different errors and uncertainties originating from these sources. First, this report provides a definition of the error or uncertainty. Then it provides best practice advice on how to avoid errors and, where this is not possible, how to estimate and reduce errors and uncertainties in the numerical solutions. These BPGs are meant to avoid—or at least reduce—user errors. User errors originate from the incorrect use of CFD and related codes because of either a lack of experience or a lack of resources. In the course of a simulation, the user may make mistakes or unwise choices that manifest themselves as one or more of the above-mentioned errors.

2

In addition, the user should be aware of the uncertainties that exist in the simulation of flow and heat transfer in dry cask applications.

The application section of this document includes one example selected from the NRC's CFD application on dry casks to show the validation process. The application example is followed by best practice advice and lessons learned from the application.

1.1   Scope

This document provides both background and guidance for the methods used to examine flows that are incompressible, steady and unsteady, laminar, and turbulent. The guidelines address viscous flow methods and the aspects of CFD common to all methods.

1.2   Structure of This Document

This document begins with a definition of the concepts of general errors and uncertainties in CFD and a comprehensive section providing guidelines on how to deal with these general errors and uncertainties. Guidelines are given to draw the user's attention to the likely sources of uncertainty when formulating a problem, and to know the sources of error inherent in CFD methods.

This is followed by a comprehensive section that deals with CFD BPGs for viscous incompressible turbulent flow calculations using Reynolds-Averaged Navier-Stokes (RANS) methods. Then, a checklist of CFD BPGs is presented. The document provides one application example using CFD to analyze and evaluate dry cask thermal response. This section also discusses many issues that are dealt with in the CFD dry cask application. As a result, it provides guidelines based on validation and sensitivity analysis.

The example in the application section is a three-dimensional (3-D) CFD model used for validation purposes. This example uses data for a ventilated storage cask (VSC-17) collected by Idaho National Laboratory (INL) to validate the CFD model. In this example, the validation process was used to remove the modeling and application uncertainties. To address the modeling uncertainties, the report focused on turbulence modeling of buoyancy-driven air flow. Similarly, in the application uncertainties, the pressure boundary conditions used to model the air inlet and outlet vents were investigated and validated.

The final section provides a brief checklist of CFD best practice guidance designed to serve as a quick reference section. This is compiled as a summary of best practice advice given in the previous sections.

2.0 GENERAL ERRORS AND UNCERTAINTIES IN CFD SIMULATIONS

2.1 Introduction

CFD is now a well-established science and is generally accepted as describing the broad topic encompassing the numerical solution, by computational methods, of the governing equations that describe fluid flow (i.e., the set of the Navier-Stokes equations, mass continuity, and additional conservation equations for such factors as heat and species concentration). The solution is done on scales down to those of the largest turbulence eddies and boundary layer widths. It is an intrinsic assumption in CFD that the details of the geometry are important to the flow and must be represented accurately. Therefore, most CFD codes use body-fitted meshes in which the faces of the mesh cells coincide with the physical boundaries of the problem (walls, inlets, outlets, etc.) For complex geometric situations, this means that very careful and time-consuming mesh generation, with mesh refinement in regions of strong gradients, is an important precursor to any complex CFD simulation. The application of CFD to complex flow problems requires considerable experience, and critical interpretation of the results must be undertaken from a position of fundamental knowledge of fluid dynamics and heat transfer. Nonetheless, the codes are only as good as the physical models programmed into them; in particular, for single-phase applications, the turbulence model must be scrutinized carefully to determine if it is appropriate to the situation being modeled. In addition, because of the complexity of modern commercial CFD packages, careful input preparation and equation solving are essential to avoiding errors.

A typical Reynolds number encountered in dry cask applications will be of the order 10^5 to 10^6. Consequently, turbulent flow conditions are to be expected. Industrial CFD simulations generally incorporate Reynolds-Averaged Navier-Stokes (RANS) turbulence models (usually the high Reynolds number k-ε model) that return only mean values for the velocities and temperatures. However, turbulence is not only a small-scale phenomenon since for the Reynolds number quoted above, the ratio of the largest to smallest turbulent eddies is 10^7 to 10^8. The RANS models average over all of these length scales to produce an estimate of the mean quantities. Most information relating to the scale of variation (turbulent flows are highly irregular and unsteady) is lost in this process, although the mean turbulent kinetic energy (k) does provide a measure of the average size of the velocity fluctuations. In addition, it has been recognized that some applications require the use of more sophisticated turbulence modeling approaches. Examples include large eddy simulation, in which the largest of the turbulence scales are computed explicitly while smaller scales are modeled or even direct numerical simulation (DNS), in which all turbulence scales down to Kolmogorov scales are computed with no modeling assumptions. Such calculations are, of necessity, 3-D and time dependent. Hence, they are very expensive computationally.

The k-ε model, although now over 30 years old, is still regarded as the industrial standard turbulence model simply because it is robust and inexpensive. This is not to say that industry is fully satisfied with the results of the model; huge amounts of extra effort are required to moderately improve predictions and, therefore, in the industrial context, these efforts are not justified. Basically, the model is one of momentum transfer (except for a few special flow types, such as impinging jet heat transfer), and extra problems with transfer occur not because of basic deficiencies in the model; instead, they result from treating turbulent heat transfer in accordance with the Reynolds analogy that relates the turbulent heat flux to the mean temperature gradient through a turbulent Prandtl number. The decision to use the Reynolds analogy is a balance between accuracy and the need for computational speed. At some point in the future, computer speeds and storage will be high enough that more detailed treatment of turbulent heat transfer will become more prevalent.

Nonetheless, the model also has rather well-known deficiencies for certain flow types (e.g., swirling flows, spreading of jets). This means that, for most CFD applications, a definite need exists to benchmark the various simulations being undertaken and to validate the code predictions against experimental data, where available.

2.2 Classical Validation Process

Today, CFD is an exact technology. However, like any precision instrument, a state-of-the-art, general-purpose, commercial CFD package is a very complex entity and demands respect in its application. The widespread use of such codes in industry and the increasing reliance now placed on predictions from the codes has prompted several recent initiatives to produce a documented "code of conduct" or "best practice guidelines" (BPGs). The objective of the present document is to provide such guidelines for dry cask applications. Nonetheless, quality assurance (QA) concerning CFD is best achieved through benchmarking and validation.

Validation examines whether the physical models used in computer simulations agree with real-world observations. The process addresses the question: Have we solved the right equations? Validation is one of the two fundamental tiers upon which the credibility of numerical simulations is built; the other is verification. The basic validation strategy is to identify and quantify both error and uncertainty through comparison of simulation results with experimental data.

2.3 Sources of Errors and Uncertainties and Their Classification

The process of performing a CFD calculation is complex and requires the CFD analyst to perform a number of different activities. These activities are listed below and typically include:

- define the problem
- select the solution strategy
- develop the computational model
- analyze and interpret the results

Each of these steps is potentially error prone or subject to some degree of uncertainty. No universally accepted means is available for identifying or classifying errors, which can range from human or user errors to inadequacies in the modeling strategy and model equations. However, the ERCOFTAC BPG adopts the following classification based on seven different sources of error and uncertainty:

(1) model error and uncertainties
(2) discretization or numerical error
(3) iteration or convergence error
(4) round-off error
(5) application uncertainties
(6) user errors
(7) code errors

This categorization has been adopted for these guidelines and, in common with the ERCOFTAC BPG, is used to structure the guidance.

2.3.1 Model Error and Uncertainty

In general, a simplified mathematical model of reality is necessary to render a simulation feasible. This is defined as modeling or uncertainty errors because of the difference between the real flow and the exact solution of the model equations. It includes errors based on the fact that the exact governing flow equations are not solved but are replaced with a physical model of the flow that may not be an accurate model of reality. In short, the model errors and uncertainties can be described as the errors that arise because we are in fact solving the wrong equations.

The most prominent example of a model error or uncertainty is the use of turbulence closure models. The Navier-Stokes equations normally are used to model the flow. However, their direct solution to describe the turbulent flow is prohibitively expensive in dry cask applications. What is known as DNS is currently restricted to flows with low Reynolds numbers in relatively simple geometries because of the very large range of scales that have to be resolved. The physical complexity of turbulent flows is reduced by using the averaged Navier-Stokes equations, in which averaging is performed in space for LES and in time for the Reynolds-averaged approach. The solution of these averaged equations, however, requires turbulence closure models that describe the influence of the unresolved scales on the resolved flow field. These approximate models then introduce errors and uncertainties to the numerical solution results.

The equations of state, relating thermodynamic variables and dependence of thermophysical properties (e.g., viscosities, diffusivities) on the variables and the data for chemical kinetics, also are examples of models that have some inherent uncertainty.

2.3.2 Discretization or Numerical Error

The spatial and temporal discretizations probably are the most crucial sources of numerical error (Casey and Wintergerste, 2000). These errors describe the difference between the exact solution of the basic system of partial differential equations and the numerical solution obtained with finite discretization in space and time. In theory, the analytical solution is approached with refinement of the discretization (increased resolution) when the discretization scheme is consistent. In general, the greater the number of grid cells, the closer the results will be to the exact solution of the modeled equations. However, both the fineness and the distribution of the grid points affect the result.

This type of error arises in all numerical methods and is related to the approximation of a continually varying parameter in space by some polynomial function for the variation across a grid cell. In first-order schemes, for example, the parameter is taken as constant across the cell. In short, discretization errors arise because we do not find an exact solution, but a numerical approximation, to the equations we are trying to solve.

Therefore, the mesh used to discretize the space is important for the accuracy of the results. Another important aspect is the approximation of the spatial and temporal variation of the flow variables. This is normally done by a polynomial representation of the variation, which then serves to define the order of a numerical approximation with regard to the truncation error of a Taylor series expansion.

2.3.3   Iteration or Convergence Error

The nonlinear algebraic system that results from the discretization of the basic system of partial differential equations is either entirely or partly solved with an iterative method or by time integration toward a steady state. These types of errors arise because of the difference between a fully converged solution on a finite number of grid points and a solution that is not fully converged. The equations that CFD methods solve start from an initial approximation to the flow solution and iterate to a final result. This should ideally satisfy the imposed boundary conditions and the equations in each grid cell and globally over the whole domain. However, if the iterative process is incomplete, then errors arise if the iteration or time integration is stopped too early. Thus, the iterative convergence error is the difference between this intermediate solution and the exact solution of the algebraic system of equations.

Judgment of the iterative convergence normally is based on the residuals, which indicate how far the present solution is away from the exact solution within each cell. The monitoring of the residuals is based on scaled norms of the residual vector for each conservation equation. In addition to the residual, the physical target values normally are monitored with the iteration number or with time. Convergence can then also be checked by requiring that these values become constant with the iteration number or with time.

In short, convergence errors arise because we are impatient, short of time, or the numerical methods are inadequate and do not allow the solution algorithm to complete its progression to the final converged solution.

2.3.4   Round-Off Errors

Round-off errors result from the computer's finite representation of numbers. Single precision numbers are stored in 32 bits and, for example, have a relative precision of 6 to 7 decimal places in FORTRAN 95. Double precision numbers use 64 bits of storage and have a relative precision of 14 to 15 decimal places in FORTRAN 95. Commercial CFD codes normally are available as single and double precision. These types of errors arise because of the fact that the difference between two values of a parameter during some iterative scheme is below the machine accuracy of the computer. This is caused by the limited number of computer digits available for storage of a given physical value.

2.3.5   Application Uncertainties

The computational domain normally contains only part of the entire control volume and environment. Therefore, the choice of the position of the boundaries of the computational domain influences the results. This influence adds to the uncertainty of the simulation results, but it can also lead to errors if the choice is inadequate. The influence of the external surroundings on the flow and heat exchange within the computational domain is taken into account with the prescription of the behavior of the flow variables at the boundaries. For the boundaries through which the flow enters the computational domain, complete information on all flow variables is necessary.

In many cases, complete information about boundary conditions is not provided. Therefore, the necessary approximation of some or all flow variables at the inflow boundaries adds to the uncertainty of the numerical results. This is also the case for the choice of the boundary conditions at solid walls. In these cases, the prescribed roughness and selected wall functions are important. Another source of error and uncertainty within this context is the simplification of the geometrical complexity present in the case. To reduce the computational costs, geometric details often are omitted. In addition, uncertainties about whether the flow is likely to be steady or unsteady can introduce large error in the simulation.

In short, application uncertainties are inaccuracies that arise because the application is complex and precise data needed for the simulation is not available.

2.3.6 User Errors

User errors are defined as errors that arise because of the mistakes and oversights of the user. Such errors generally decrease with increasing experience of the user but cannot be completely eliminated because "to err is human." The popular expression "garbage in, garbage out" often describes this error.

2.3.7 Code Errors

Code errors pertain to CFD codes in general and therefore to the code developers' area of responsibility. Computer programming or software errors are mistakes that exist in the computerized model. Errors made while programming the conceptual model can be detected and removed with code verification. Other errors can originate from the use of the code on different platforms (e.g., hardware, operating systems, compilers, run-time libraries). These errors are treated within the realm of software quality engineering (Oberkampf et al., 2004). Neither type of errors falls under the responsibility of the code users.

Such errors often are difficult to find because CFD software is highly complex and typically involves hundreds of thousands of lines of code for a commercial product. Computers are very unforgiving, and even a relatively simple typing error that might easily be overlooked (such as an "i" for a "j" in a single word) when incorporated into a single line of code can have disastrous consequences.

2.4 Definitions of Errors and Uncertainties

The deficiencies or inaccuracies of CFD simulations can be related to a wide variety of errors and uncertainties. A publication of the American Institute of Aeronautics and Astronautics (AIAA), "AIAA Guide for the Verification and Validation of Computational Fluid Dynamics Simulations" (AIAA, 1998), provides useful definitions of error and uncertainty in CFD as follows:

- Error: A recognizable deficiency not caused by a lack of knowledge.
- Uncertainty: A potential deficiency caused by a lack of knowledge.

These philosophical definitions can be made clearer by examples. Typically known errors are round-off errors in a digital computer and the convergence error in an iterative numerical scheme. In these cases, the CFD analyst has a reasonable chance of estimating the likely magnitude of the error. Unacknowledged errors include mistakes, either in the input data or in the implementation of the code itself, and no methods are available to estimate their magnitude. Uncertainties arise because of incomplete knowledge of a physical characteristic, such as the turbulence structure at the inlet to a flow domain or because uncertainty exists in the validity of a particular flow model being used. An error is something that can be removed with appropriate care, effort, and resources. An uncertainty, however, cannot be removed because it is rooted in a lack of knowledge.

2.5   Definitions of Verification, Validation, and Calibration

In discussions of CFD errors and uncertainties, the distinction in meaning between the terms validation, verification, and calibration is useful. The definitions used in these guidelines closely follow similar definitions given in the AIAA guide (1998), Roache (1998), Rizzi and Voss (1998), and Fisher and Rhodes (1996):

- Verification: Procedure to ensure that the program solves the equations correctly.

- Validation: Procedure to test the extent to which the model accurately represents reality.

- Calibration: Procedure to assess the ability of a CFD code to predict global quantities of interest for specific geometries of engineering design interest.

3.0 CFD BEST PRACTICE GUIDELINES

Best practice guidelines (BPGs) provide procedures for the model user to estimate and reduce errors and uncertainties in the results of a numerical simulation. There are several initiatives to establish CFD BPGs. For industrial CFD in general, the ERCOFTAC BPGs exist (Casey and Wintergerste, 2000) and provide valuable information on general topics of CFD that are also relevant to dry cask applications. The ERCOFTAC guidelines also focus on the industrial end user of CFD codes and not on the evaluation and validation of CFD codes. Best practice guidelines on CFD for engineering problems also have been published by the Thematic Network for Quality and Trust in the Industrial Application of CFD (QNET) that was handed over to the control of ERCOFTAC. The OECD/NEA/CSNI CFD working groups also provided practical guidance for application of single-phase CFD to the analysis of nuclear reactor safety (NEA/CSNI/R(2007)5, 2007). Two other references consulted were BPGs for marine applications of CFD (MARNET-CFD) (WS Atkins, 2002) and European Cooperation in Science and Technology (COST) (Franke et al, 2007). These two references provided comprehensive BPGs for the marine industry and the urban environment, respectively. The ASME Standard V&V 20 addresses verification and validation of CFD. The purpose of V&V 20 is to assess the accuracy of CFD simulation.

The BPGs presented in this document are based on the works cited above, as well as the example shown in the application section of this report. These guidelines mainly are intended to solve RANS equations. Only general guidelines are extracted from the above references since most parameters depend largely on the details of the application problem.

This document structures the guidelines according to the general steps of conducting a numerical simulation. These general steps are listed below and include:

- geometry and grid generation
- discretization errors
- convergence errors
- modeling uncertainty
- round-off errors
- user errors
- code errors

3.1 Grid Requirements

The computational grid is a discretized representation of the geometry of interest. It should provide an adequate resolution of the geometry and the expected flow features. The grid's cells should be arranged in a way to minimize discretization errors. Specific recommendations here generally follow those provided by the Evaluation of Computational Fluid Dynamic Methods for Reactor Safety Analysis (ECORA) (Menter, 2002) and ERCOFTAC (Casey and Wintergerste, 2000).

3.1.1 Geometry Generation

Before the grid generation can start, the geometry has to be created or imported from computer-aided design (CAD) data or other geometry representations. Attention should be given to:

- Use of correct coordinate systems.

- Use of correct units.

- Completeness of the geometry. If local geometrical features with dimensions below the local mesh size are not included in the geometrical model (e.g., fuel element assemblies), they should be incorporated through a suitable empirical model.

- Oversimplification because of physical assumptions. Problems can arise, for example, when the geometry is oversimplified or when symmetry conditions are used that are not appropriate for the physical situation.

- Location of boundary conditions. The extent of the computational domain has to capture relevant flow and geometrical features. A major problem can be the positioning of boundary conditions in regions of large gradients or geometry changes. If in doubt, the sensitivity of the calculation to the choice of computational domain should be checked.

When the geometry is imported from CAD data, these data should be checked thoroughly. Frequently, CAD data have to be adapted (cleaned) before they can be used for mesh generation. For instance, some mesh generators require closed 3-D volumes (solids) for mesh generation, and these are not always directly obtained from CAD data. Consequently, the CAD data have to be modified. However, care must be taken to ensure that these changes to the geometry do not influence the computed flow.

3.1.2 Geometrical Uncertainties

In many industrial and engineering problems, the geometry of the object to be simulated is extremely complex and requires much effort to specify it exactly for a computer simulation. Many sources of error can arise in this process, such as the following:

- Changes in geometry that occurred during the design process.

- The CAD geometry definition is insufficiently complete for flow simulation. Some surfaces and curves may not meet at the intended endpoint locations because of different levels of accuracy in different parts of the CAD model. Other curves may be duplicated.

- The geometry of a tested component may get modified during the testing procedure, and these modifications may not have been added to the original drawings.

- The geometry may not be manufactured within the tolerances as shown on the drawing, particularly for fine flow features, such as grid spacers, or symmetrical features.

- The effective geometry of the surface may have changed during use because of wear, erosion, or fouling.

- Small details of the geometry may have been omitted (e.g., roughness on the walls, welding fillet radii, small protrusions from the body).

- The coordinate system used in the CAD system may differ from what is used in the CFD code (rotational direction).

Guidelines

- Check and document that the geometry of the object being calculated is the geometry as intended. For example, the transfer of geometrical data from a CAD system to a CFD system may involve loss of surface representation accuracy. Visual display of the geometry helps.

- Generally, it is not necessary to explicitly include geometric features that have dimensions below that of the local grid size, provided they are taken into account in the modeling (e.g., roughness in wall layer).

- In areas where local detail is needed, grid refinement in local areas with fine details should be used, such as in the neighborhood of fine edges or small clearance gaps. If grid refinement is used, the additional grid points should lie on the original geometry and not simply be a linear interpolation of more grid points on the coarse grid.

- Check that the geometry is defined in the correct coordinate system and with the correct units requested by the CFD code. CAD systems often define the geometry in millimeters and this must be converted to SI units if the code assumes that the geometry information is in meters. This is commonly done by most codes.

- If the geometry is altered or deformed by the hydrodynamic, mechanical, or thermal loading, then some structural or mechanical calculation may be necessary to determine the exact geometry.

3.1.3 Grid Design

In a CFD analysis, the flow domain is subdivided into a large number of elements or control volumes. In each computational cell, the model equations are solved, yielding discrete distributions of mass, momentum, and energy. The number of cells in the mesh should be sufficiently large to obtain an adequate resolution of the flow geometry and the flow phenomena in the domain. As the number of elements is proportional to storage requirements and computing time, many 3-D problems require a compromise between the desired accuracy of the numerical result and the available number of cells. The available cells need to be distributed in a way that minimizes discretization errors. This leads to the use of nonuniform grids, hybrid grids consisting of different element types, overset grids, and local grid refinement. Modern CFD methods use body-fitted grids in which the cell surfaces follow the curved solution domain. Different mesh topologies can be used for this purpose as follows:

- Structured grids (which consist of hexahedral elements). Cell edges form continuous mesh lines that start and end on opposite block faces. The control volumes are addressed by a triple of indices (i, j, k). The connectivity to adjacent cells is identified by these indices. Hexahedral elements theoretically are the most efficient elements and are well suited for the resolution of shear layers. The disadvantage of structured grids is that they do not adapt well to complex geometries, although this problem can be eliminated through the use of an overset (overlapping structured) grid.

- Unstructured grids. Meshes are allowed to be assembled cell by cell freely without considering continuity of mesh lines. Hence, the connectivity information for each cell face needs to be stored in a table. The most typical cell shape is the tetrahedron, but any other form, including hexahedral cells, is possible. This results in an increase of storage requirements and calculation time.

- Hybrid grids (which combine different element types, i.e., tetrahedral, hexahedra, prisms, and pyramids). This grid combines structured with unstructured meshes. The grid must be fine enough to capture all important flow features, which may be achieved by local grid refinement. Unstructured meshes especially are well suited for this purpose. If block structured grids are used, then local refinement results in block attachments with a dissimilar number of grid lines. Some CFD codes provide algorithms to adapt the grid resolution locally according to numerical criteria from the flow solution, such as gradient information or error estimators. The accuracy of the simulation increases with increasing number of cells (i.e., with decreasing cell size). However, because of limitations imposed by increased computer storage and runtime, some compromise is nearly always inevitable. In addition to grid density, the quality of a mesh depends on various criteria, such as the shape of the cells (e.g., aspect ratio, skewness, included angle of adjacent faces), distances of cell faces from boundaries, or spatial distribution of cell sizes. The introduction of special topological features, such as O-grids or C-grids and care taken to locate block-interfaces in a sensible manner, can help improve the overall quality of a block-structured mesh. Unstructured meshing techniques may take advantage of prism layers with structured submeshes close to domain boundaries.

- Block structured grids. For the sake of flexibility, the mesh is assembled from a number of structured blocks attached to each other. Attachments may be regular (i.e., cell faces of adjacent blocks match) or arbitrary (i.e., general attachment without matching cell faces).

3.1.4 Choice of the Computational Grid

When referring to the computational grid, one must first define the discretization method that will be used for the basic equations. The following discussion is restricted to the closely related finite volume and finite difference methods, with a strong bias toward the finite volume method since it is widely used in commercial CFD codes. For the finite element discretization method, different requirements exist for the quality of the computational grid (Casey and Wintergerste, 2000). With the finite volume and finite difference methods, the computational results crucially depend upon the grid used to discretize the computational domain. The grid has to be designed so that it does not introduce errors that are too large. This means that the resolution of the grid should be fine enough to capture the important physical phenomena like shear layers and vortices with sufficient resolution. Also, the quality of the grid should be high. Ideally, the grid is equidistant.

Therefore, grid stretching and compression should be small in regions of high gradients to keep the truncation error small. The expansion ratio between two consecutive cells should be below 1.3 in these regions. Scaperdas and Gilham (2004) and Bartzis et al. (2004) recommend a maximum of 1.2 for the expansion ratio. However, higher-order numerical schemes might allow larger changes as the absolute value of the truncation error is smaller than with lower-order schemes (Schroeder et al., 2006). For LES, nonequidistant grids correspond to nonuniform filter widths. Their application to the basic equations leads to filtered equations that contain more unknown terms than the subgrid stresses. This is because filtering with a nonuniform filter width and a partial spatial derivation do not commute and therefore are called commutation errors. They describe the change in the definition of the resolved and subgrid flow variables because of the varying filter width. Ghosal and Moin (1995) have shown that the commutation error terms are second-order in the filter width. When using numerical approximations of the second-order in the grid size, these commutation error terms are therefore normally neglected.

For the widely used finite volume methods, another criterion for grid quality is the angle between the normal vector of a cell surface and the line connecting the midpoints of the neighboring cells (Ferziger and Perić, 2002). Ideally, these should be parallel. As to the shape of the computational cells, hexahedra are preferable to tetrahedral since the former are known to introduce smaller truncation errors and display better iterative convergence (Hirsch et al., 2002). On walls, the grid lines should be perpendicular to the wall (Casey and Wintergerste, 2000); (Menter, 2002). Therefore, if a tetrahedral grid is to be used, prismatic cells should be used at the wall with tetrahedral cells away from the wall. For example, Fothergill et al. (2002) found improved results for a prismatic/tetrahedral grid when compared to a purely tetrahedral grid.

Recommendations for the grid resolution generally cannot be made in advance because this is highly problem dependent. If simulations employ the logarithmic wall model, the position of the first computational node should, of course, be placed in the logarithmic region, corresponding to a nondimensional wall distance of at least 30 (Casey and Wintergerste, 2000). For typically very rough walls, this is automatically satisfied (Hargreaves and Wright, 2006). However, for validation simulations, a systematic grid convergence study using generalized Richardson extrapolation should be tried. This is straightforward for CFD codes using the RANS approach, which does not require grid-dependent parameterizations. For the Richardson extrapolation, solutions on at least three systematically refined/coarsened grids are necessary. From these simulation results, the error band (uncertainty) of the spatial discretization error of the solution on the finest mesh can be estimated. Details on the method are provided in Appendix A.

When a global systematic grid refinement is not possible because of resource limitations, then a local grid refinement should be used at least in the areas of main interest.

Most commercial CFD codes can perform local refinement using the local gradient or curvature of the flow variables. The choice of these indicator functions should depend on the target variables, which will be compared with experimental data.

3.1.5 Grid Quality

A good mesh quality is essential for performing a good CFD analysis. Therefore, assessment of the mesh quality before performing large and complex CFD analyses is very important. Most mesh generators and CFD solvers check the mesh parameters, including grid angles, aspect ratios, face warpage, right-handedness, negative volumes, etc. The CFD user should check the guide of the applied mesh generators and CFD solver for specific requirements. General recommendations for generating high-quality grids are as follows:

- Clean up CAD geometry. For body-fitted grids, check that the surface grid conforms to the CAD geometry.

- Avoid highly skewed cells, in particular for hexahedral cells or prisms in which the included angles between the grid lines should be optimized so that the angles are about 90 degrees. Angles with less than 40 degrees or more than 140 degrees often show a deterioration in the results or lead to numerical instabilities, especially in the case of transient simulations.

- Ensure that the angle between the grid lines and the boundary of the computational domain (the wall or the inlet- and outlet-boundaries) is close to 90 degrees. This recommendation is stronger than the recommendation that the angles in the flow field far away from the domain boundaries.

- Avoid jumps in grid density. Growth factors between adjacent volumes should be smaller than 2.

- Avoid grid lines not aligned with the flow direction (e.g., tetrahedral meshes, in thin wall boundary layers). Computational cells not aligned with the flow direction can lead to significantly larger discretization errors.

- Avoid the use of tetrahedral elements in boundary layers. Away from boundaries, ensure that the aspect ratio (the ratio of the sides of the elements) is not too large. This aspect ratio typically should not be larger than 20. Near walls, this restriction may be relaxed and can be beneficial.

- Observe the code requirements of mesh stretching or expansion ratios (rates of change of cell size for adjacent cells). The change in mesh spacing should be continuous and mesh size discontinuities should be avoided, particularly in regions of high gradients.

- Ensure that the mesh is finer in critical regions with high-flow gradients, such as regions with high shear and where there are significant changes in geometry or where suggested by error estimators. Make use of local refinement of the mesh in these regions according to the selected turbulence wall modeling.

- Check the assumption of regions of high-flow gradients assumed for the grid with the result of the computation and rearrange grid points if it is found to be necessary.

- Analyze the suitability of the mesh by a grid dependency study (this could be local) in which at least three different grid resolutions are used. If this is not feasible, try to compare different order of spatial discretizations on the same mesh.

- Use a finer and more regular grid in critical regions (e.g., regions with high gradients or large changes, such as free surfaces).

- Avoid the presence of nonmatching grid interfaces in critical regions. An arbitrary grid interface occurs when no one-to-one correspondence exists between the cell faces on both sides of a common geometry face.

- Use the local grid refinement in areas where local details are needed to capture fine geometric details. If grid refinement is used, the additional grid points should lie on the original boundary geometry and not simply be a linear interpolation of more grid points on the original coarse grid.

If the target variables of a turbulent flow simulation include wall values, such as wall heat fluxes or wall temperatures, the choice of the wall model and the corresponding grid resolution can have a large effect on the results. Typical wall functions are:

- calculation of the wall shear stresses and wall heat fluxes based on logarithmic velocity and temperature profiles

- calculation of the wall shear stresses and wall heat fluxes based on linear velocity and temperature profiles

- calculation of the wall shear stresses and wall heat fluxes based on linear or logarithmic velocity and temperature profiles

Wall functions of this kind are used for all RANS turbulence models and also for large eddy simulations (LESs) and detached eddy simulations (DESs). The choice of the wall model has a direct influence on the mesh design. The following values are recommended for the distance of the first grid point away from the wall:

- Logarithmic wall functions: $30 < y^+ < 500$. The upper limit is Reynolds number-dependent. The limit decreases for decreasing Reynolds numbers. A logarithmic near-wall region does not exist for very small Reynolds numbers.

- Linear wall functions: $y^+ < 5$. Linear wall functions only can be used in combination with special low Reynolds versions of the k-ε turbulence model; k-ω-type models usually do not need special modifications.

- Linear or logarithmic wall functions: $y^+ < 500$. Linear or logarithmic wall functions can only be used in combination with special low Reynolds versions of the k-ε turbulence model; k-ω-type models usually do not need special modifications.

Here y^+ is the nondimensional wall distance:

$$y^+ = \frac{\rho u_\tau y}{\mu} = \frac{\sqrt{\rho \tau_w} y}{\mu}$$

The recommendations above are strictly only valid for attached flows. The logarithmic law is not valid for separated flows. Close to separation, the wall shear stress τ_w goes to zero, and with it the nondimensional wall distance y^+, irrespective of the physical wall distance, y. In the above equation, ρ and μ are the density and viscosity of the fluid. In contrast, the linear near-wall law remains valid but requires finer resolution. The combination of logarithmic and linear wall functions yields the best generality and robustness against small variations of the near-wall grid distance. For two-dimensional (2-D) flows, the following equation is valid:

$$u_\tau = U_e \sqrt{\frac{c_f}{2}}$$

U_e is the velocity at the boundary layer edge or a characteristic reference velocity. The skin friction coefficient c_f for turbulent flows is typically in the interval from 0.003–0.005. With these two values, the friction velocity u_τ and the distance of the first grid point away from the wall can be a priori estimated as:

$$y = \frac{y^+ \mu}{\rho u_\tau}$$

17

Finally, this report makes the following recommendations on the choice of element types:

- Hex elements are the most efficient elements from a numerical point of view. They require the least memory and computing time per element. They can be well adapted to shear layers (long and thin), for instance, in the vicinity of walls. However, generation of hex meshes in complex geometries often requires a large manual and cognitive effort.

- If this effort seems too high, the use of tetrahedral meshes is a viable alternative. Tetrahedral elements require roughly 50 percent more memory and computing time per element than hex elements. They are not very efficient for the resolution of shear layers. Either a large number of tetrahedral elements must be used or the grid angles become very small. If wall values are the target values of a calculation, pure tetrahedral meshes should be avoided or used with great care.

- The combination of tetrahedral elements in the flow domain and prism elements close to walls is a reasonable alternative to the use of pure tetrahedral grids. The combination of tetrahedral elements in the flow domain and hex elements close to walls (with pyramids as transition elements) is a better alternative than pure tetrahedral grids.

- Nonmatching grid interfaces, which combine different grid types and mesh densities, should be avoided if possible. They can have a negative effect on accuracy, robustness (convergence), and parallel scalability (depending on the numerical algorithm and the application).

Based on these observations, the following rules and priorities can be formulated to obtain the best accuracy and efficiency:

- Use pure hex element grids, if the grid generation effort is manageable.

- Use hybrid grids with hex elements close to the wall and tetrahedral elements in the core of the domain.

- Use hybrid grids with prism elements close to walls and tetrahedral elements in the core of the domain.

- Use pure tetrahedral element grids.

The order becomes reversed if the manual grid generation effort is the sorting parameter. The user decides which grid to use. However, the final documentation of the analysis should include the reasoning that has led to the use of a particular grid and topology.

A grid dependence and sensitivity study always should be performed to analyze the suitability of the mesh and to provide an estimate of the numerical error of the results. At least two (three is better) grids with significantly different mesh sizes should be used. If this is not feasible, results obtained with different discretization schemes in time and space can be compared on the same mesh.

3.2   Discretization Schemes

Ideally, selection of discretization schemes should be automated within the CFD code and not be a user option. Unfortunately, the current state of CFD presents the user with a list of potential discretization schemes with some general advice on situations in which each is appropriate. Selecting temporal and spatial discretization is a balancing act between too much numerical diffusion for low-order schemes, and spatial wiggles (unphysical nonmonotonic behavior) in key state variables with higher-order schemes.

Hirt (1967) quantified the concept of numerical diffusion for first-order numerical schemes. Consider a simple one-dimensional (1-D) advection equation, approximated with backward Euler time (fully implicit) and first-order upwind spatial discretization. Applying Hirt's analysis, the numerical solution can be shown to closely approximate the analytic solution of an advection-diffusion equation as shown below:

$$\frac{\partial \rho}{\partial t} + \frac{\partial}{\partial x}(\rho V) = D\frac{\partial^2 \rho}{\partial x^2},$$

where the numerical diffusion coefficient D is

$$D = \frac{V}{2}(V\Delta t + \Delta x)$$

An analyst contemplating numerical methods that are first-order accurate in time or space should obtain typical values for turbulent diffusion coefficients (or molecular diffusion coefficients if the flow is laminar) and use the previous formula to estimate the time step and mesh size needed to make the numerical diffusion substantially less than the physical diffusion. In cases in which physical diffusion is unimportant to a problem, numerical diffusion should at least be limited to the point that it does not significantly distort the results of advection terms.

In general, use of first-order discretization should be avoided. The one significant exception is steady flow solutions. In some cases, a CFD code will be unable to converge to its steady-state iteration when using an appropriate higher-order spatial discretization. In this situation, an initial steady solution usually can be obtained with a first-order spatial method, and then this is used as a starting point for iteration to steady state with the higher-order method. However, even this approach does not always work, and the CFD code may be indicating that vortex shedding is significant and no steady solution exists.

Higher-order methods remove second derivative terms from Taylor truncation error analysis that give rise to obvious numerical diffusion. However, they do not completely suppress numerical diffusion.

The analyst's problem is quantifying the magnitude of numerical diffusion relative to turbulent diffusion in a given simulation. The Richardson-based error analysis described in Appendix A is a way to determine that errors introduced by numerical diffusion are bounded. Higher-order upwind methods typically are selected for use in RANS calculations. However, LES, DES, and DNS calculations need the lower numerical diffusion associated with central difference methods (typically fourth order or higher). For methods operating on a logically rectangular mesh, performance is optimal when flow is aligned with a mesh direction. Results should be studied with particular care when flow is diagonal to the mesh lines. All higher-order methods have the potential for cell-to-cell spatial oscillations in key state variables, and results, particularly near continuity or shock waves, should be watched carefully for this behavior. When these oscillations are severe, they can be controlled by a flux correction method (available in any serious CFD code). Such techniques automatically are applied to limited areas and reduce the spatial accuracy to first order in these regions.

When evaluating tests of discretization schemes, it is important to keep a proper prospective. Understand that the results of a Richardson error analysis probably will indicate lower effective order of accuracy than advertised for the selected discretization scheme. The important goals are to demonstrate convergence of the solution as the mesh or time step is refined and to achieve acceptably low numerical distortion of important physical phenomena at the discretization used in the final analysis.

3.2.1 Spatial Discretization Schemes

Different numerical methods evaluate the fluxes at the same grid locations as the transported quantities or somewhere in between (collocated or staggered grids). In both cases, an algebraic approximation of the spatial functions is required to calculate the gradients at these locations. This approximation is called the differencing scheme in finite volume or difference methods or the basis function in finite element methods. The accuracy of the scheme depends on the form of the algebraic relationship and on the number of grid points used. The spatial discretization or truncation error equals the difference between the scheme and the exact formulation based on a Taylor expansion series. A formally second-order scheme is consistent with the exact formulation, up to the term with a power of 2. A third-order scheme also takes into account the next higher term. The formal order of accuracy is not preserved on irregular meshes where it reduces by 1. Reducing the cell size by introducing a finer grid has the greatest impact on the accuracy of the solution if higher-order schemes are applied. Halving the elements in all directions using a third-order scheme will reduce the numerical error by a factor of 8, while this factor is only 2 with a first-order scheme.

If the solution of the physical problem exhibits only small gradients, even a first-order scheme may be acceptable. However, this option is not at all suitable for general engineering applications involving complex flows with large gradients and thin boundary layers. The large truncation error introduced by the first-order upwind scheme, particularly popular in finite volume methods, is known as numerical viscosity or diffusivity because it gives rise to artificial diffusion fluxes that may be much stronger than the real molecular or turbulent contributions. On the other hand, higher-order schemes suffer from a different, more obvious problem: namely, the appearance of a characteristic wavy pattern with a wavelength of two cell sizes in the neighborhood of steep gradients. Dispersion errors cause these so-called wiggles (i.e., waves with different wave lengths are not transported with the same speed). Dispersion errors are most prominent in central differencing schemes for finite volume methods and quadratic basis function schemes for finite element methods. Higher-order upwind schemes are less prone to this problem.

Guidelines

- Avoid the use of first-order upwind schemes. The use of methods of higher-order (at least second) is recommended for all transported quantities. It may be necessary to use a first-order scheme at the start of a calculation, as it is likely to be more robust; however, as convergence is approached, a second-order or higher scheme should be used.

- Try to give an approximation of the numerical error in the simulation by applying a mesh refinement study or, if this is not possible, by mesh coarsening.

- Make use of the calculation of an error estimator (which may be based on residuals or on the difference between two solutions of different order of accuracy), if it is available in the code.

3.2.2 Time Discretization Schemes

Purely steady flow fields with the time-derivative equal to zero are only a special case of the time-dependent equations. Generally, fluid flows are transient, whereby the sources for this time-dependent behavior are as follows:

- external transient or nontransient forces

- transient boundary conditions, moving walls (e.g., the fluttering of an airfoil)

- vortex stretching, a 3-D phenomenon caused by the nonlinear term of the governing equations that also leads to the fluctuating nature of turbulence

The computation of steady turbulent flow is the most common kind of simulation for the general use of CFD. In these cases, the Reynolds-averaged flow is steady, while the average turbulent quantities account for the time-dependence of the turbulent fluctuations. However, the RANS-equations also allow the time-dependent Reynolds-averaged flow fields to be computed, based on the assumption that global unsteadiness does not affect the temporal average of the turbulent quantities. This is physically correct if the spatial scale of the turbulent eddies is much smaller than the geometrical scale of the analyzed geometry. A time-dependent simulation is required if the scale of eddies or vortices becomes larger and is comparable in size to the dimensions of the geometry (e.g., the computation of vortex shedding).

If an accurate spatial discretization is applied, physically time dependent flows will fail to converge using a steady-state method. Convergence problems with a steady simulation very often can be interpreted as a hint that the flow is unsteady and a time-stepping scheme would be appropriate. On the other hand, symmetry boundary conditions may impose a steady flow, although in reality it would be transient. If the complete geometry, including both sides of the symmetry plane, were used, the velocity field would oscillate perpetually. Averaging the solution over a long time interval would lead to a symmetrical field, however, which would differ from the steady-state solution with the symmetry plane.

The temporal discretization scheme provides an approximation of the time derivative. Most CFD codes offer first-order and second-order schemes that are unconditionally stable and most effective in terms of computer memory and stability requirements. Low-storage, higher-order Runge-Kutta methods also are available. The order of the scheme and the choice of the time step influence the size of the amplitude and the phase error—the two components of the temporal discretization error. To improve time accuracy, self-adaptive time-stepping procedures (such as predictor-corrector methods) can be used.

The choice of the time step depends on the time scales of the flow being analyzed. If time steps are too large, the simulation might fail to capture important flow features and mimic unphysical steady behavior. It is therefore advisable to start with relatively small Courant-Friedrichs-Lewy (CFL) numbers, even though this is not required from the point of view of numerical stability. The CFL number for incompressible flow is defined as $CFL = v\,\Delta t/\Delta x$, in which Δt is the time step, Δx is the local cell size and v the local velocity. Some CFD codes use a time-stepping scheme for steady-state simulations. It should be noted that the accuracy of the converged steady-state result is not completely independent of the time step. Special care is required to avoid choosing a time step that is too large.

Guidelines

The overall solution accuracy is determined by the lower-order component of the discretization. At least second-order accuracy is recommended in space and time. For time-dependent flows, the time and space discretization errors are strongly coupled. Therefore, finer grids or higher-order schemes are required (in both space and time).

- Check the influence of the order of the temporal discretization by analysis of the frequency and time-development of a quantity of interest (e.g., the velocity in the main flow direction).

- Check the influence of the time-step on the results.

- Ensure that the time step is adapted to the choice of the grid and the requested temporal size by resolving the frequency of the realistic flow and ensure that it complies with eventual stability requirements.

The highest frequency should be resolved with at least 10 to 20 time steps per period (Menter, 2002). Another method to estimate the time step in advection-dominated problems is the relation $\Delta t = CFL\ \Delta x_{min}/U_{max}$, in which Δx_{min} is the minimum grid size, U_{max} is the maximum velocity, and CFL is the Courant-Friedrichs-Lewy number. Choosing the minimum grid spacing and the maximum velocity makes this estimate conservative. In several models, the time step is determined continuously as the minimum of all time steps calculated per grid point.

To assess the influence of the time-step size on the results, a systematic reduction or increase of the time step should be made and the simulation repeated. The two results then can be analyzed with the Richardson extrapolation as described in Appendix A.

3.3   Convergence Control

Convergence is a major issue with the use of CFD software. Fluid mechanics is involved with nonlinear processes, dealing with inherently unstable phenomena, such as turbulence. CFD software simulates these physical processes and therefore is subject to the same issues as the processes it is trying to represent. As such, it is not guaranteed that there will be a steady-state "converged" solution to a problem.

CFD problems generally are nonlinear, and the solution techniques use an iterative process to successively improve a solution until "convergence" is reached. Convergence can mean many things to different people. More formally, in mathematics, convergence describes limiting behavior, particularly of an infinite sequence or series toward some limit. To assert convergence is to claim the existence of a limit, which may be unknown. For any fixed standard of accuracy, you can always be sure to be within the limit, provided you have gone far enough.

As this definition indicates, the exact solution to the iterative problem is unknown, but you want to be sufficiently close to the solution for a particular required level of accuracy. Convergence, therefore, does need to be associated with a requirement for a particular level of accuracy. This requirement depends on the purpose to which the solution will be applied.

Convergence also is often measured by the level of residuals, the amount by which discretized equations are not satisfied, and not by the error in the solution. Therefore, the user should be aware of this when deciding what convergence criterion should be used to assess a solution.

The following describes two meanings of convergence that are in common use in CFD. Both forms of convergence shall be checked to understand the accuracy of a calculation.

3.3.1 Differential Versus Discretized Equations

The first convergence refers to the formal process that brings the exact solution of the discretized equation set ever closer to the exact solution of the underlying partial differential equations as each of the discretization sizes for independent variables approaches zero. That is:

$$T_j^n \to \overline{T}\!\left(x_j, t_n\right) \text{ as } \Delta x_j, \Delta t \to 0$$

In practice, the definition is not very useful because exact solutions of algebraic equations (with no round-off errors) generally are difficult to obtain, and exact solutions of the partial differential equations are even more so, except for a few oversimplified demonstration cases. However, in the case of linear equations, the concept of convergence can be linked with consistency and stability, which are easier to demonstrate.

A system of algebraic equations generated by a space and time discretization process is said to be consistent with the partial differential equation if, in the limit of the grid spacing and the time step tending to zero, the algebraic equation is identical with the partial differential equation (PDE) at each grid point at all times. Consistency may be demonstrated by expressing the differences appearing in the discretized equations in terms of Taylor expansions in space and time and then collecting terms. For consistency, the resulting expression will be identical with the underlying PDE, apart from a set of remainder terms that should all tend to zero as Δx_j, $\Delta t \to 0$. In CFD, almost universally, the numerical schemes for solving the fluid flow and energy equations are consistent simply because of the methodology used in their development.

Numerical stability, however, is far more difficult to prove, and most formal procedures are limited to linear equations. In a strict sense, stability only applies to marching problems (i.e., to the solution of hyperbolic or parabolic equations), and will be defined here accordingly. A numerical scheme is considered to be stable if errors arising from any source (e.g., round-off or truncation) do not grow from one time step to the next. The most common example of instability arises from the use of explicit time-differencing for convective problems in which the time step exceeds the CFL criterion. Physically, this corresponds to information being numerically transported within a time step faster than the physical communication process, either by sonic or fluid velocities. In practical terms, small disturbances grow until the solution is destroyed. Classical methods are available for determining the stability of numerical schemes, but most of the work refers to linear systems.

The Lax Equivalence Theorem states that, given a well-posed, linear, initial-value problem (well-posed means the solution develops in a continuous manner from the initial conditions) and a finite difference approximation to it that satisfies the consistency condition, stability is a necessary and sufficient condition for convergence of the numerical result to the analytic solution as discretization is refined. The theorem is very powerful because, as noted, it is much easier to demonstrate consistency and stability than convergence directly, although convergence is the most useful property in the sense of quality and trust in the solution. Although the theorem is stated in terms of finite differences, it also applies to other discretization schemes, such as finite volume and finite element. The theorem can only be rigorously applied to linear, initial-value problems, whereas with CFD the governing equations are nonlinear and of the boundary- or mixed-initial- or boundary-value type. In these circumstances, the Lax Equivalence Theorem should be regarded as a necessary, but not sufficient, condition and used heuristically to provide a pragmatic solution strategy (i.e., one that is consistent and stable).

Although users have no guarantee of convergence to the solution of the Navier-Stokes differential equations, they should use common sense to look for obvious signs of trouble. Frequently, analysts assume that step-to-step oscillations associated with bounded numerical instabilities are oscillating about the correct mean solution to the problem. This may not be the case, and isolated time-step sensitivity studies should be performed on any such case to determine shift in mean behavior with time-step size.

Although convergence of results as time-step or mesh size is reduced toward zero does not guarantee that the numerical solution is converging to the solution of the set of PDEs, it is a good indicator. If no convergence can be seen in these sensitivity analyses, convergence to the PDE solution will not occur.

3.3.2 Termination of Iterative Solvers

The second meaning of convergence refers to the criterion adopted to terminate an iterative process. Such processes nearly always come up in CFD simulations because of (1) implicit or semi-implicit time differencing, and (2) the nonlinear nature of the governing equations.

For a fully coupled solver, all the governing equations are considered part of a single system and solved together. All variables are updated simultaneously, and there is just one overall iteration loop. For highly nonlinear equations in 3-D that occur in industrial CFD applications, this requires a large memory overhead. Until recently, such approaches were considered impractical. However, with the advent of large-memory machines and fast central processing units (CPUs), the approach has become tractable, and modern commercial CFD software is built around the concept of fully-coupled solvers.

An alternative is to treat each of the governing equations in isolation, assuming all other variables are fixed, and to invert the subsystem matrix on this basis. This procedure is often called the inner iteration. The other equations are then solved in turn, repeating the cycle, or outer iteration, until all the equations are satisfied simultaneously.

The solution of the fully-coupled system of equations, and the inner loop of the noncoupled system, requires the solution of a set of linear, simultaneous equations; in other words, the inversion of a matrix. Except for small problems for which inversion by Gaussian elimination can be attempted, the solution algorithm is usually iterative. In fact, the success of finite-volume discretization schemes in CFD is largely because the algorithms produce diagonally dominant system matrices. Such matrices can be readily inverted using iterative methods. Many such methods have been derived, ranging from the classical Jacobi, Gauss-Seidel, successive-over-relaxation, and alternative direction implicit algorithms, through the more modern Krylov family of algorithms (e.g., conjugate-gradient or the more up-to-date multigrid and algebraic multigrid methods.) All such methods involve pivoting on the diagonal entry for each row of the matrix, and the success and speed of convergence of the iteration process essentially is governed by how much this term dominates over the sum total of the others in the row (supported by under-relaxation, if necessary) and the accuracy of the initial guess.

When using iterative solvers, it is important to know when to stop and examine the solution (steady-state problems) or to move on to the next time step (transient solutions). The difference between two successive iterates, measured by an appropriate norm, being less than a preselected value, is not sufficient evidence for solution convergence, but the information may be used to provide a proper estimate of the convergence error as follows. The largest eigenvalue (or spectral radius), λ_m, of the iteration matrix may be estimated from the (rms or L_2) norms at successive iteration steps according to the following:

$$\lambda_m \approx \left\| r^n \right\| / \left\| r^{n-1} \right\| \text{ where } r^n = \Phi^{n+1} - \Phi^n; \; \Phi \text{ is a dependent variable, and } n \text{ the iteration number.}$$

Regardless of the underlying iteration scheme, CFD users should perform some simple numerical studies to understand the effect of convergence criteria on solution accuracy. After a base run, a second run should be performed with all iterative convergence criteria halved. After plotting results for key variables, the user can make a practical decision on significance of the discrepancies. To make a conservative judgment of impact, all differences in results should be doubled.

3.3.3 Convergence Criteria

In general, setting convergence criteria is difficult because it depends on the user's requirements and the software package used. For example, for a simple demonstration calculation, a not-well-converged solution might be adequate to illustrate some principles and for use as a starting point for a more refined calculation. In other situations, in which the quantity of interest is in a small localized area of the flow and the flow rates there are much smaller than inlet flow rates (e.g., heat transfer inside a small separated region), a very high level of convergence might be required.

Users should consider what they want to achieve and use a number of different criteria to assess whether a solution is converged. These criteria include the residuals given by the software package being used; global imbalances in mass, momentum, energy, etc.; whether key global quantities, such as heat transfer from or to a surface forces on a body, have reached an equilibrium value; and whether information from solution monitor points has stabilized. These monitor points should be in areas where the flow could be much weaker and not where the flow could be converged easily (e.g., just downstream of an inlet).

3.3.4 Choice of Iterative Convergence Criteria

Most computer programs use iterative methods to solve the algebraic system of equations (e.g., the equation for the pressure). Starting from an initial guess, the flow variables are recalculated in each of the iterations until the equations are solved up to a user-specified error. The termination criterion usually is based on the residuals of the corresponding equations. These residuals should tend toward zero. Scaling of the residuals is usually done with the residuals after the first iteration. The scaled residual then shows how much the initial error has dropped. In industrial applications, typically a termination criterion of 0.001 is used, which generally is too high to have a converged solution. A reduction of the residuals of at least four orders of magnitude is recommended. For validation purposes of turbulence or other physical models, much lower criteria should be used. If the residual is driven down to its theoretical value of machine accuracy (10^{-10}–10^{-12} for double precision), no more iterative error is present in the solution. In addition to the residuals, the target variables also should be recorded. If these variables are constant or oscillate around a constant value, then the solution can be regarded as converged. The same should be done for the integral balances of mass, momentum, and energy. Based on the behavior of the target variables and the integral balances, it can be decided which termination criterion for the residuals is sufficient. A quasi-constant behavior of these values can be expected if stationary solutions are sought (VDI, 2005). The values may change in correspondence to changes in boundary values or other source and sink terms for unsteady runs. This procedure also should be followed when unsteady simulations will be performed. Implicit time integration methods require iterations within the time steps, so the above should be applied within each time step.

3.3.5 Convergence Errors

Iterative algorithms are used for steady-state solution methods and for procedures to obtain an accurate intermediate solution at a given time step in transient methods. Progressively better estimates of the solution are generated as the iteration count proceeds. There are no universally accepted criteria for judging the final convergence of a simulation, and mathematicians have found no formal proof that a converged solution for the Navier-Stokes equations exists. In some situations, the iterative procedure does not converge—it either diverges or remains at a fixed and unacceptable level of error or oscillates between alternative solutions. Careful selection and optimization of control parameters (such as damping and relaxation factors or time steps) may be needed in these cases to ensure that a converged solution can be found.

The level of convergence is most commonly evaluated based on residuals; on values of globally integrated parameters, such as lift coefficient or heat-transfer coefficient; or on time and iteration signals of a physical quantity at a monitor point that is an arbitrarily selected location in the flow domain.

3.3.6 Residuals

Residuals are 3-D fields associated with a conservation law, such as conservation of mass or momentum or energy. They indicate how far the present approximate solution is away from perfect conservation (balance of fluxes). Usually, the residuals are normalized by dividing by a reference value, which may be one of the following:

- maximum value of the related conserved quantity
- average value of the related conserved quantity
- inlet flow of a related quantity

Convergence usually is monitored on the basis of one representative number characterizing the residual level in the 3-D flow field. This single value may be:

- a maximum value
- the sum of absolute values
- the sum of squared values
- the arithmetical average of absolute values
- the root-mean-square value

26

The large number of variants makes it difficult to give precise statements on how to judge convergence and at which residual level a solution may be considered converged. In principle, a solution is converged if the level of computer round-off error is reached. Special care is needed in defining equivalent levels of convergence if different codes are used for comparison purposes.

Recommendations to the code developers include:

- CFD codes should make available the maximum possible information to judge convergence. This includes residuals for every conserved quantity.

- Information on the spatial distribution of residuals should be provided.

- Residuals should be dimensionless.

- The handbook should include a clear definition of how the residuals are determined.

- One commonly accepted definition of the residual should be adopted to avoid confusion for CFD users.

Recommendations for the users include:

- Be aware that different codes have different definitions of residuals.

- Always check the convergence on global balances (conservation of mass, momentum, energy, and turbulent kinetic energy) where possible, such as the mass flow balance at inlet and outlet and at intermediate planes within the flow domain.

- Check not only the residual but also the rate of change of the residual with increasing iteration count.

- Do not assess convergence of a simulation purely in terms of the achievement of a particular level of residual error. Carefully define solution-sensitive target quantities for the integrated global parameters of interest and select an acceptable level of convergence based on the rate of change of these (e.g., mass flow, heat transfer, and moment forces on a body).

- Perform a test for each application of the effect of converging to different levels of residual on the integrated parameter of interest (this can be a single calculation that is stopped and restarted at different residual levels). This test demonstrates at what level of residual the parameter of interest can be considered to have converged and identifies the level of residual that should be aimed for in similar simulations of this application.

- Monitor the solution in at least one point in a sensitive area to see if the region has reached convergence.

- For calculations proving difficult to converge, consider the following advice:

  o Use more robust numerical schemes during the first (transient) period of convergence and switch to more accurate numerical schemes as convergence improves.

- Modify parameters controlling convergence (e.g., under relaxation parameters or the CFL number).

- If the solution is heavily under-relaxed, increase relaxation factors at the end to see if the solution holds.

- Check if switching from a steady to a time-accurate calculation has any effect.

- Consider using a different initial condition for the calculation.

- Check the numerical and physical suitability of boundary conditions.

- Check if the grid quality in areas with large residual has any effect on the convergence rate.

- Look at the residual distribution and associated flow field for possible hints (e.g., regions with large residuals or unrealistic velocity levels).

3.4   Modeling Uncertainty

3.4.1   Solution Algorithm

The discretized set of RANS equations can be solved with various solution procedures, such as pressure-based or density-based methods (for a review, see Ferziger and Perić (1999), Fletcher (1991), or Hirsch (1991)). The solution algorithms use numerous tuning parameters, such as artificial time steps, under-relaxation, etc., to improve convergence behavior and robustness of the code. The code's field of application and the modeling technique included influence the choice of the numerical method and the solution procedure. In principle, the solution of a well-converged simulation is independent of the numerical method and the solution algorithm selected.

Guidelines

Check the adequacy of the solution procedure with respect to the physical properties of the flow. As a first step in this process, the parameters controlling convergence (e.g., relaxation parameters or Courant number) of the solution algorithm should be used as the CFD-code vendor or developer suggests.

If parameters need to be changed to aid convergence, it is not advisable to change too many parameters in one step because it then becomes difficult to analyze which of the changes influenced the convergence. In cases of persistent divergence, see sections on boundary conditions, grid, discretization, and convergence errors. Carefully consider if the flow can be expected to exhibit a steady or unsteady flow behavior. Consider the size of the unsteady scales expected to be present in the flow field in comparison to the geometrical dimensions. If this is large, then an unsteady simulation is necessary.

If a steady solution has been computed and there is a reason to be unsure that the flow is really steady, then an unsteady simulation should be performed with the existing steady flow field as the initial condition. Examination of the time development of the physical quantities in the locations of interest will identify if the flow is steady or not.

3.4.2 Guidelines on Turbulence Modeling

The user should be aware that no universally valid general model of turbulence is available that is accurate for all classes of flows. Validation and calibration of the turbulence model is necessary for all applications.

If possible, the user should examine the effect and sensitivity of results to the turbulence model by changing the turbulence model being used. The relevance of turbulence modeling only becomes significant in CFD simulations when other sources of error, in particular the numerical and convergence errors, have been removed or properly controlled. Clearly, no proper evaluation of the merits of different turbulence models can be made unless the discretization error of the numerical algorithm is known, and grid sensitivity studies become crucial for all turbulence model computations.

3.4.3 Weaknesses of the Standard k-ε Model

Despite the great variety of turbulence modeling options available to the user, the standard k-ε model with wall functions, as set out by Launder and Spalding (1974), remains the workhorse of industrial computation. Therefore, cataloging the major weaknesses associated with this model in practical application is valuable. Furthermore, palliative actions that might be considered should be indicated when possible. These are listed below. The advisory actions are drawn from extensive literature on the subject and should not to be viewed as definitive cures. Manuals of CFD codes may offer alternative and equally effective advice, and many commercial codes include alternatives to the standard k-ε model. Where the action given below involves a modification or adjustment to the standard k-ε model, this should be regarded as specific palliative for the weakness under consideration and usually will not prove to be of general benefit (and may even be worse).

Guidelines

- The turbulent kinetic energy is over-predicted in regions of flow impingement and reattachment leading to poor prediction of the development of flow around leading edges and bluff bodies. Ince and Launder (1995) have proposed a modification to the transport equation for ε, which is designed to tackle this problem.

- Regions of recirculation in a swirling flow are underestimated. Reynolds stress models (RSM) should be used to overcome this problem.

- Highly swirling flows generally are predicted poorly because of the complex strain fields. RSM or nonlinear eddy viscosity models should be used in these cases.

- Mixing is poorly predicted in flows with strong buoyancy effects or high streamline curvature. RSM should be used in these cases.

- Flow separation from surfaces under the action of adverse pressure gradients is poorly predicted. The real flow is likely to be much closer to separation (or more separated) than the calculations suggest. The Baldwin-Lomax one-equation model is often better than the standard k-ε model in this respect (Baldwin and Lomax, 1978). The shear stress transport (SST) version of Menter's k-ω based, near-wall resolved model also offers a considerable improvement (Menter, 1993, 1996).

- Flow recovery following reattachment is poorly predicted. Avoid the use of wall functions in these regions.

- The spreading rates of wakes and round jets are predicted incorrectly. The use of nonlinear k-ε models should be investigated for these problems.

- Turbulence-driven secondary flows in straight ducts of noncircular cross section are not predicted at all. Linear eddy viscosity models cannot capture this feature. Use RSM or nonlinear eddy viscosity modeling.

- Laminar and transitional regions of flow cannot be modeled with the standard k-ε model. This is an active area of research in turbulence modeling. No simple practical advice can be offered other than advocating user intervention to switch the turbulence model on or off at predetermined locations. Another option is to use low Reynolds k-ε as well as the transitional SST version of Menter's k-ω based.

3.4.4   Near-Wall Modeling

In wall-attached boundary layers, the normal gradients in the flow variables become extremely large as wall distance reduces to zero. Many mesh points packed close to the wall are required to resolve these gradients. Moreover, as the wall is approached, turbulent fluctuations are suppressed and, eventually, viscous effects become important in the region, which is known as the viscous sublayer. This modified turbulence structure means that many standard turbulence models are not valid all the way through to the wall. Thus, special wall-modeling procedures are required.

3.4.4.1   Wall Functions

The difficult near-wall region is not explicitly resolved with the numerical model but is bridged using so-called wall functions (Rodi, 1981) and (Wilcox, 1998). To construct these functions, the region close to the wall is characterized in terms of variables rendered dimensionless with respect to conditions at the wall. The wall friction velocity u_τ is defined as

$$u_\tau = \left(\frac{\tau_w}{\rho}\right)^{1/2}$$

Where τ_w is the wall shear stress. Let y be normal distance from the wall and let U be time-averaged velocity parallel to the wall. Then the dimensionless velocity U^+ and dimensionless wall distance y^+ are defined as

$$\frac{u}{u_\tau} \text{ and } \frac{\rho y u_\tau}{\mu} \text{ respectively.}$$

If the flow close to the wall is determined by conditions at the wall, then U^+ can be expected to be a universal function of y^+ up to some limiting value of y^+. This is observed in practice, with a linear relationship between U^+ and y^+ in the viscous sublayer and a logarithmic relationship, known as the law of the wall, in the layers adjacent to this (so-called "log-layer"). The y^+ limit of validity depends on external factors, such as pressure gradient and the penetration of far field influences. In some circumstances, the range of validity also may be affected by local influences, such as buoyancy forces, if there is strong heat transfer at the wall. The turbulence velocity ($k^{1/2}$) and length scales, when treated in the same way, also exhibit a universal behavior.

These universal functions can be used to relate flow variables at the first computational mesh point, displaced some distance y from the wall, directly to the wall shear stress without resolving the structure in between. The only constraint on the value of y is that y^+ at the mesh point remains within the limit of validity of the wall functions. A similar universal, nondimensional function can be constructed that relates the temperature difference between the wall and the mesh point to heat flux at the wall (Rodi, 1981). This can be used to bridge the near-wall region when solving the energy equation.

Wall Function Guidelines

- The meshing should be arranged so that the values of y^+ at all the wall adjacent mesh points are greater than 30 (the form usually assumed for the wall function is not valid much below this value). It is advisable that the y^+ values do not exceed 100 and should certainly never be less than 11. Some commercial CFD codes account for this by switching to alternative functions if y^+ is <30. Be aware of this and check user manuals.

- Cell-centered schemes have their integration points at different locations in a mesh cell than cell-vertex schemes. Thus, the y^+ value associated with a wall adjacent cell differs according to which scheme is used on the mesh. Care should be exercised when calculating the flow using different schemes or codes with wall functions on the same mesh.

- The values of y^+ at the wall adjacent cells strongly influence the prediction of friction and, hence, drag. Therefore, particular care should be given to the placement of near-wall meshing if these are important elements of the solution.

- Check that the correct form of the wall function is being used to take into account the wall roughness.

3.4.4.2   Near-Wall Resolution

As already mentioned, a universal near-wall behavior over a practical range of y^+ may not be realizable everywhere in a flow. Under such circumstances, the wall-function concept breaks down and its use will lead to significant error, particularly if wall friction and heat-transfer rates are important. The alternative is to fully resolve the flow structure all the way to the wall. Some turbulence models can be validly used for this purpose; others cannot. For example, the k-ω two-equation model can be deployed through to the wall, as can the one-equation k-L model (Wolfshtein, 1969), but the standard k-ε and RSM models cannot. Various so-called low-Reynolds number versions of the k-ε and RSM models have been proposed incorporating modifications that remove this limitation (Patel et al., 1985) and (Wilcox, 1998). Alternatively, the standard k-ε and RSM models can be used in the interior of the flow and coupled to the k-L model used to resolve just the wall region. This is known as a two-layer model.

Whatever modeling approach is adopted, a large number of mesh points must be packed into a very narrow region adjacent to the wall to capture the variation in the flow variables.

Near-Wall Resolution Guidelines

- Ensure that the turbulence model being used is capable of resolving the flow structure through to the wall.

- Ensure that the value of y^+ at the first node adjacent to the wall is close to unity.

- Use a small stretching factor for progressing the mesh spacing away from the wall. There should be at least 10 mesh points between the wall and y^+ equal to 20.

3.5 Round-Off Errors

Round-off errors can be significant for high-Reynolds number flows where the boundary layer resolution can lead to very small cells near the wall. The number of digits of a single precision simulation can be insufficient for such cases. The only way to avoid round-off errors with a given CFD code is to use a double precision version. In case of erratic behavior of the CFD method, the use of a double precision version is recommended.

Round-off errors usually are not of great significance. However, in situations where the small arithmetical differences between two large numbers become relevant, cancellation because of round-off may lead to severe errors. To avoid large values, it is common practice to calculate pressure relative to a reference value. Examples in which round-off errors are known to be of significance are as follows:

- low-Reynolds number turbulence models with large exponential terms
- flows with density-driven buoyant forces with small density and temperature differences
- high aspect ratio grids with large-area ratios on different sides of the grid
- conjugate heat transfer
- calculations of scalar diffusion with low concentrations of one species
- low Mach number flows with a density-based solver
- flows with large hydrostatic pressure gradients

Guidelines

- Always use the 64-bit representation of real numbers (double precision on common UNIX workstations).

- Developers are recommended to use the 64-bit representation of real numbers (REAL*8 in FORTRAN) as the default settings for their CFD code.

3.6 User Errors

3.6.1 General Comments

In CFD, human factors play an important role because results largely depend on the competence and expertise of the user. It is worth discussing this aspect of CFD because it is one of the prime causes of uncertainty in CFD simulation results. This discussion may help to avoid some, if not all, of the most easily avoidable mistakes. User errors result from the inadequate use of the resources available for a CFD simulation. The resources are, for instance, the problem description, computing power, CFD software, physical models in the software, and the project timeframe. Several factors may result in user errors:

- Lack of attention to detail, sloppiness, carelessness, and mistakes.

- Optimistic and uncritical use of CFD, thanks to the high accessibility of commercial software with simple interactive graphical user interfaces and the convincing and seductive power of the colorful visualizations.

- Lack of experience so that the user is unaware of a technical difficulty or that critical information is missing.

- Unfamiliarity with a particular CFD code and the tacit assumption that certain parameter settings are equivalent to those in a code with which the user is more familiar.

- Oversimplification of a given problem (e.g., geometry, equation system).

- Poor geometry and grid generation.

- Use of incorrect boundary conditions.

- Selection of nonoptimal physical models.

- Incorrect or inadequate solver parameters (e.g., time step).

- Acceptance of nonconverged solutions and postprocessing errors.

3.6.2 Control of the Working Process

Many mistakes are made by lack of attention to detail, or because the user is not aware of factors that can lead to them. One way to deal with these issues is for the user to have a checklist of issues that can arise to ensure that all relevant problem areas have been dealt with. This becomes most important if the user has limited experience.

A formal management QA system with checklists can support the inexperienced user to produce quality CFD simulations. Roache (1998) has noted, however, that a CFD project can meet all formal QA requirements and still be of low quality (or flatly erroneous). On the other hand, high-quality work can be done without a formal QA system.

The guidelines below provide examples of the types of issues that should be dealt with in a formal QA management system. The issues covered are based on the process of performing a CFD simulation.

3.6.3 Guidelines on Problem Definition

The user should give careful thought to the requirements and objectives of the simulation and typically consider the following points:

- Is the prescribed CFD simulation method appropriate (e.g., for buoyancy-driven problems, is the RANS approach most appropriate)?

- Are the objectives of the simulation clearly defined?

- What are the accuracy requirements?

- What local or global quantities are needed from the simulation?

- What are the documentation and reporting requirements?

- What are the important flow physics involved (e.g., steady, unsteady, single phase, laminar, turbulent, transitional, internal, external)?

- What is the area of primary interest (domain) for the flow calculation?

- Is the geometry well-defined?

- What level of validation is necessary? Is this a routine application in which validation and calibration already have been performed on similar flow fields and only relatively small changes can be expected from earlier similar simulations? Or, is it a nonroutine application, in which little earlier validation work has been done?

- What level of computational resources is needed for the simulation (e.g., memory, disk space, CPU time), and are these available?

3.6.4 Transient or Steady Model

The choice between transient and steady state is only an issue with RANS-based simulations. More detailed simulations based on LES, DES, and DNS fundamentally are transients. Most selections are based on common sense. The only serious problems in making the choice arise in configurations that appear steady based on imposed boundary conditions but may be shedding vortices or contain fundamentally unstable macroscopic flow patterns.

One option for questionable flows is to run a transient and inspect the flow patterns. If the user wants to start the analysis running a CFD code in steady-state mode, he or she should understand the code's algorithm for obtaining steady state. If the specific CFD code achieves steady-state solutions through some pseudotransient iteration procedure, it generally will not converge if the flow is fundamentally transient. However, if the algorithm is a direct solution of flow equations with no time derivative terms, it may provide a smooth answer that masks actual transient behavior.

3.6.5 Laminar and Transition Flows

The distinction between laminar, transitional, and turbulent flow is difficult. Sometimes the flow appears in different states depending on the location of the area of interest (e.g., the flow in an inlet of a dry cask can be laminar, while the flow inside the cask is transitional or turbulent). The general problem of the transition from laminar to turbulent flow and the computation of the origin of turbulence is a subject of fundamental academic research. It cannot be included in general industrial CFD computations. The simplest way around this problem is to calculate the flow as a turbulent one. The turbulent kinetic energy is about zero in the nominal laminar flow regimes. Special care needs to be taken if a turbulence model with wall function is used to obtain information about wall shear stress.

34

Guidelines

- Check that the flow does not contain extensive regions of laminar or transition flow that the k-ε turbulence model would estimate incorrectly.

3.6.6 Choice of Initial Data

In RANS, URANS (unsteady Reynolds-Averaged Navier-Stokes), and LES models, a boundary and initial value problem has to be numerically solved. The larger the model domain, the more relevant the initial data become. For RANS, stationary solutions are searched, thus the iteration is stopped as soon as the solution is no longer changing or the solution converges. In these cases, the boundary values mainly influence the model solution, and the impact of the initial data is small. Initializing with a flow field that is close to the final solution will reduce the computational efforts necessary to reach stationary solutions.

For URANS and LES, the initial data determine the time-dependent development in the beginning of the simulation. As a rule of thumb, the impact time can be estimated with a relation that includes the domain size and environmental conditions. During this initial period, the model results are very dependent on the initial data and should not be interpreted as a solution, which reflects the final flow.

Initial data and inflow data very often are used as one and the same. This is a good starting point for most models. However, if these initial data (and therefore the inflow profiles) do not correspond to the situation to be investigated (e.g., wrong wind direction), then a model result comparable to the situation to be modeled cannot be expected. Because initial data are not known perfectly but include uncertainties that result from measurement inaccuracy or a lack of representativeness of the measurement site, the initial input values are never perfectly known.

The initial data uncertainty should be reduced as much as possible by evaluating the reliability of the initial data and choosing only those initial data with small uncertainties. However, quite often the number of input data is not even sufficient to know all variables that need to be initialized. There are rarely several data to choose from and the input data uncertainty generally is unknown. In all these (common) cases, the uncertainty of the input data should be estimated (e.g., from other experiments or from experience). Sensitivity studies in the uncertainty range of the initial data (e.g., for different inflow directions) allow for estimating the effect of initial data uncertainty on model results. This can be a very costly effort, and currently no method is established to determine which sensitivity studies are most worthwhile to perform to derive the information on the initial data influence on model results. The resulting probability distribution for the model results currently can only be calculated when using a massive amount of computer resources. Therefore, the current best practice advice is to keep initial data uncertainty as little as possible and to keep in mind that the initial data influence the model results in unsteady simulations.

3.6.7 Choice of Boundary Conditions

Boundary conditions represent the influence of the surroundings that the computational domain has cut off. Since boundary conditions largely determine the solution inside the computational domain, their proper choice is very important. Often, however, these boundary conditions are not fully known. Therefore, the boundaries of the computational domain should be far enough away from the region of interest. This will prevent solution contamination with the values applied at the boundaries.

Two types of boundary conditions and combinations of them are most commonly encountered. The Dirichlet condition specifies the distribution of a physical quantity over the boundary at a given time step. The Neumann condition defines the distribution of its first derivative. Users normally have no control over the spatial discretization in the neighborhood of boundaries. The CFD code developer should ensure that the boundary region retains the overall accuracy of the numerical scheme. There is common consent that a good practice for outflow boundaries is to set the convective derivative normal to the boundary face equal to zero and to combine this with a streamwise extrapolation of transported quantities. The same treatment usually is applied at pressure boundaries. Open boundaries cause the following difficulties:

- nonphysical reflection of outgoing information back into the domain

- difficulties in providing information about the properties of the fluid that may inadvertently enter the domain from the outside

- difficulties that arise if open boundary conditions are placed in regions of high swirl, large curvatures, or pressure gradients. Some CFD codes prevent fluid from entering into the domain through open boundaries. To avoid undesirable side effects, open boundaries should be placed very carefully.

Guidelines

- Ensure that appropriate boundary conditions are available for the case under consideration. For swirling flows, consult the manual to ensure appropriate boundary condition are used (e.g., radial equilibrium of pressure field instead of constant static pressure). Special nonreflecting boundary conditions sometimes are required for outflow and inflow boundaries where there are strong pressure gradients (Giles, 1990).

- Check if the CFD code allows inflow at open boundary conditions. If inflow cannot be avoided at an open boundary, then ensure that the transported properties of the incoming fluid, including turbulence boundary conditions, are properly modeled.

3.6.8 Application of Boundary Conditions

In many real applications, frequently there is difficulty defining some boundary conditions at the inlet and outlet of a calculation domain in the detail necessary for an accurate simulation. A typical example is the specification of the turbulence properties (turbulence intensity and length scale) at the inlet flow boundary because these are practically arbitrary in dry cask CFD. The user should be aware of these problems and develop a good feel for the certainty or uncertainty of the imposed boundary conditions. This can be achieved best if the user knows and understands the application being analyzed. Additional uncertainties can arise because boundary condition data that need to be specified are inconsistent with the model used.

General Guidelines on Boundary Conditions

- Examine the possibilities of moving the domain boundaries to a position where the boundary conditions are more readily identified, well-posed, and can be precisely specified.

• For each application, an uncertainty analysis should be performed in which the boundary conditions are systematically changed within certain limits to see the variation in results. Should any of these variations prove to have a sensitive effect on the simulated results and lead to large changes in the simulation, then it is clearly necessary to obtain more accurate data on the specified boundary conditions.

3.6.9 Inflow Boundary Conditions

The use of a turbulence model (other than an algebraic model) requires the specification of turbulence properties at a domain inlet region. Verified quantities should be used as inlet boundary conditions for k and ε, because the magnitude can significantly influence the results. If no data are available, then some sensitivity calculations should be performed to examine the influence of the selected values.

Guidelines on Inlet Conditions

• Examine the possibilities of moving the domain inlet boundaries to a position where the boundary conditions are easily identified, well-posed, and can be precisely specified.

• Perform a sensitivity analysis for each application where the inlet boundary conditions are systematically changed within certain limits. Aspects that should be examined include the following:

    o   inlet flow direction and magnitude

    o   uniform inlet velocity (slug flow) or velocity profile

    o   variation of physical parameters

    o   variation of turbulence properties at inlet (see below)

3.6.10 Guidelines on Specification of Turbulence Quantities at an Inlet

A particularly important issue is the specification of the turbulence properties at the inlet to the computational domain. Verified quantities should be used as the inlet boundary for turbulent kinetic energy k and dissipation ε, if these are available, since the magnitude can influence the results significantly.

If no data are available, then the values need to be specified using sensible engineering assumptions, and the influence of the choice should be examined by performing a sensitivity study.

For the specification of the turbulent kinetic energy k, values appropriate to the application should be used. These values generally are specified through a turbulence intensity level. ERCOFTAC guidelines suggest a variety of values depending on flow type. In hydrodynamics, low "inlet" turbulence levels are likely, but zero turbulence will bring about anomalies in turbulence modeling unless specialized approaches to laminar and transitional regions are adopted.

The specification of the turbulent length scale, as an equivalent parameter for the dissipation ε, is more difficult. For external flows, a value determined from the assumption that the ratio of turbulent and molecular viscosity μ_t/μ is of the order of 10 is appropriate. For simulations in which the near-wall region is modeled (e.g., in two-layer modeling of boundary layers), the length scale should be based on the distance to the wall and be consistent with the code's internal modeling.

If more sophisticated distributions of k and ε are used, they need to be consistent with the velocity profile so that the production and dissipation term in the turbulence equations are in balance. An inconsistent formulation (such as a constant velocity profile and constant profile of turbulence intensity at the inlet) leads to an immediate unrealistic reduction of the turbulence quantities after the inlet. These can be checked by making a plot of the ratio of turbulent to molecular viscosity μ_t/μ in cases in which problems arise. The inflow boundary should be moved sufficiently far away from the region of interest so that an inlet boundary layer can develop.

For RSM models, the stresses need to be specified and, as these normally are not available, an assumption of isotropic flow conditions with zero shear stresses is generally made.

3.6.11 Guidelines on Outlet Conditions

- The boundary conditions imposed at the outlet should be selected to have a weak influence on the upstream flow. Extreme care is needed when specifying flow velocities and directions on the outlet plane. The most suitable outflow conditions are weak formulations involving specification of static pressure at the outlet plane.

- Particular care should be taken in strongly swirling flows in which the pressure distribution on the outlet boundary is strongly influenced by the swirl and cannot be specified independently of the swirl coming from upstream.

- Be aware of the possibility of inlet flow inadvertently occurring at the outflow boundary, which may lead to difficulties in obtaining a stable solution or even to an incorrect solution. If it is not possible to avoid this by relocating the position of the outlet boundary in the domain, then one possibility to avoid this problem is to restrict the flow area at the outlet, provided that the outflow boundary is not near the region of interest.

- If there are multiple outlets, either pressure boundary conditions or mass flow specifications can be imposed depending on the known quantities.

3.6.12 Guidelines on Solid Walls

- Care should be taken that the boundary conditions imposed on solid walls are consistent with both the physical and numerical models used.

- If roughness on the wall is not negligible, significant levels of uncertainty can arise through incorrect specification of roughness within the wall function. When no detailed information is available, great care is needed.

3.6.13 Steady Flow, Symmetry, and Periodicity

A symmetric steady computation or a computation with periodic boundary conditions often is performed to reduce the computing time and required memory. There are many applications in which the nominal geometry is symmetric but the flow is asymmetric, and the flow field can be asymmetric even in the case of perfect symmetry of the geometry (e.g., an oblate spheroid at very high incidence). This can be an important factor in predicting the detail of the dynamic behavior of fluid flows. The Reynolds number is the main parameter that gives a preview of the symmetrical behavior. If the Reynolds number is high, the flow tends to be asymmetric. This asymmetry also can be forced if the real inflow conditions are not geometrically perfectly symmetrical or some distortions are within the inlet flow.

Because of the physical temporal instability of the flow (e.g., the von Karman vortex-street) or time-dependent boundary conditions, the flow field can be unsteady. This effect should be carefully examined because the flow solvers often can compute a spurious steady solution of the flow field that contradicts the physics. In cases with very strong unsteady effects within the flow field, the solution algorithm does not always converge to a steady solution.

Guidelines

- Symmetry and periodicity planes assume that the gradients perpendicular to the plane are either zero (for symmetry) or determined from the flow field (periodicity). If symmetry or periodicity planes cross the inlet or outlet boundaries, then care should be taken to specify inlet or outlet variables that are consistent with these.

- Check carefully whether the geometry is symmetric or if a geometrical distortion or disturbance in the inlet conditions is present that can trigger asymmetric solutions.

- Estimate the Reynolds number of the inflow and check if the flow could be asymmetric, turbulent, or unsteady (Roache, 1998).

- After obtaining a steady solution, switch to the transient mode and check if the solution remains stable.

- If there are difficulties to get a converged steady solution, especially if there is an oscillation of the residuals, switch to the transient mode.

- In case of doubt, the simulation should be unsteady and without symmetry assumption at the boundaries.

3.6.14 Guidelines on Solution Strategy

- Having established a clear problem definition, the user shall translate this into a solution strategy involving issues and questions that have been addressed in earlier chapters of this document, such as the following:

  o   mathematical and physical models

  o   pressure or density-based (coupled) solution method

  o   turbulence model

  o   available code/solver

  o   computational mesh

  o   boundary conditions

3.6.15 Guidelines on Code-Handling

- A potential source of user errors is in implementing the solution strategy with a particular code. Such errors might be minimized by using a formal checklist or by letting another CFD analyst check through the code input data. The following questions should be considered:

  - Have the boundary conditions not only been properly defined, but also properly applied?

  - Has the appropriate system of units been used?

  - Is the geometry correct?

  - Are the correct physical properties specified?

  - Have the intended physical and mathematical models been used (e.g., gravity forces, rotation, user-defined functions)?

  - Have default parameters been changed that may affect the solution?

  - Has the appropriate convergence criterion been defined and used?

3.6.16 Guidelines on Interpretation

- Be mindful of presentation: the solution may not be correct just because it has converged and produced high-quality color plots (or video presentations) of the CFD simulations. Ensure that an elementary interpretation of the flow-field explains the fluid behavior and that the trends of the flow analysis can be reconciled with a simple view of the flow.

- Ensure that the mean values of engineering parameters derived from the simulation are computed consistently (e.g., mass-averaged values, area-averaged values, time-averaged values). Calculation of local and mean engineering parameters with external postprocessing software may be inconsistent with the solution method of the code used (e.g., calculating shear stresses from the velocities, calculating shear stresses using nodal values instead of wall functions). Check that any test data used for comparison with the simulations also are computed in the same way as the data from the simulation.

- Consider whether the interpretation of the results and any decisions made are within the accuracy of your computation.

3.6.17 Guidelines on Documentation

- Keep accurate records of the simulation with clear documentation of assumptions, approximations, simplifications, geometry, and data sources.

- Organize the documentation of the calculations so that another CFD expert can follow what has been done.

- Be aware that the level of documentation required strongly depends on the analysis requirements as defined in the problem definition.

3.6.18 Training Requirements for CFD Users

The growth in the use of CFD codes and the trend for them to become rich packages with many alternative modeling options steadily increases the risk of user errors. This trend is reinforced by the ease of use of modern computer codes with simple graphical user interfaces, making them available for inexperienced users. Although efforts are taken to simplify the use of CFD codes, careful training with realistic exercises still should be considered the starting point for any CFD user. The theoretical part of the training should focus on fundamental modeling features, their underlying assumptions, and their limitations. The same information also is a central part of good user documentation. Unlike linear finite element stress analysis, CFD still requires expertly trained users for accurate results. In situations where inexperienced users have to be used, some restriction on their freedom to adjust critical parameters might be advisable, and they should be limited to simulations of routine types. Depending on the CFD software, additional training on grid generation is advisable.

Guidelines

- A CFD user for nonroutine applications should have good training and knowledge in classical fluid mechanics, a broad understanding of numerical methods, and detailed knowledge of the application being examined. As such, they will be able to understand the limitations of the models used (e.g., turbulence, radiation, buoyancy-driven flows).

- Training and education requirements for more routine applications can be less stringent, provided that clear guidelines or procedures have been established for the code being used. An example of a routine application would be the simulation of a standard component in a design environment in which many previous designs have been calculated and only relatively small changes in geometry and boundary conditions occur.

- In both routine and nonroutine applications, training on the use of the specific CFD code with the solution of realistic exercises is needed.

3.7   Code Errors

A software error is defined as an inconsistency in the software package. This includes the code, its documentation, and technical support. Software errors occur when the information provided to the user on the model equations is different from the actual equations that the code solves. This difference can be a result of: coding errors (bugs), deficiencies in the numerical algorithms, errors in the graphical user interface, documentation errors, and incorrect support information. Many software errors can be detected by the verification tests described previously. However, software vendors should ensure the functionality of the software through a systematic program of quality control, including extensive testing. If more than one software package meets a user's modeling needs, it is worth reviewing the quality control procedures for each candidate before making a final selection.

The success of a code generally leads to it becoming more widely used. As the user base expands, there are increasing demands for more options and the code becomes more complex. As the code deals with more difficult problems, there is again an expansion of its use. In the end, it is inevitable that code errors will be discovered by the many users who outnumber the developers by an order of magnitude and have a much wider range of applications and test cases than the code developers.

The size and complexity of large CFD software packages inevitably mean that code errors (bugs) still may be present in the software, even if it has been in use and development for many years. The painstaking but straightforward process of verification provides a way to check that the code faithfully reproduces the model approximations incorporated in the algorithms being programmed. The main problem associated with code verification is that the accuracy of a code can never be formally demonstrated for all possible conditions and applications and for all possible combinations of valid code input options

Guidelines for the Code Developer and Vendor

- The code developer or vendor should demonstrate that he or she has applied stringent methods of quality control to the software development and maintenance.

- The code developer or vendor should perform the code verification, and he or she should provide the necessary information on the verification process to the user.

- The code developer or vendor should maintain and publish a databank of verification test cases used for testing. The cases should include simple code-verifications tests (e.g., obtained solutions are independent of coordinate systems).

- The code developer or vendor should provide documentary evidence of the verification tests that the software has undergone, including clear details of the code options used during testing.

- The code developer and vendor should repeat a standard set of verification test cases for all new versions of the code.

- The code developer and vendor should supply a list of known bugs and errors in each version of the code (e.g., hotline, password-secured Web page.) This list should demonstrate that the number of bugs reduces as the code matures.

- The code developer should try to include warning notices and guidance for the user in the output (e.g., when basic rules on grid generation (expansion ratios, skew, etc.) are being broken, when important specific default options are being overruled by the code input data, or when the near wall grid is inconsistent with the turbulence modeling).

Guidelines for the Code User

- The user should recognize that codes can only be validated and verified for a class of problems involving specific variables. If the user is moving into an area in which the code is not fully verified, there is a greater risk of code errors.

- A suite of test cases set up and run by the user on new code releases provides an independent check on the code and highlights changes between releases (e.g., in default parameters).

- When a code error is suspected, the user should communicate this to the code vendor or developer as soon as possible, especially if no list of known bugs has been published. Other users may then profit from this experience or the user may find that the bug is well known and a solution or workaround is available.

- In communication with the code developer or code vendor about a suspected program error, the user should provide a concise description of the problem and all the necessary input data files so that the error can be reproduced. In cases in which commercial sensitivity precludes this, special arrangements will have to be made.

4.0   ANALYSIS OF RESULTS, SENSITIVITY STUDIES, AND UNCERTAINTIES

4.1   Analysis of Results

Most commercial CFD codes come with some kind of postprocessing package that allows many of the flow phenomena to be visualized or plotted in graphical form. The main purpose for postprocessing the results is to determine the following:

- whether the results are sensible
- whether the results are accurate

Checking the believability of the solution may involve several steps, such as checks on conserved variables, visual confirmation that velocities and pressures are smoothly distributed, and comparison with other similar problem results. The convergence history will offer some indication of whether the problem has reached a steady-state solution.

Guidelines

- Check conserved variables, including an overall heat/mass balance.

- Check that temperatures, velocities, forces, pressures, etc., have realistic values.

- Check if fluid variables, such as temperature, velocity, and pressure, are smoothly distributed over the body and vary rapidly only where expected. Discontinuities may be the result of insufficient mesh.

- Perform some simple hand calculations to check orders of magnitudes of variables.

- Run simple versions of the problem (e.g., with reduced geometry) to get an idea of the numbers involved. The accuracy of the result can only truly be determined by knowing the answer in advance. Since this is rarely the case, the accuracy of the solution will depend on the validation and suitability of the code, the approximations made, the quality of the input parameters, and the independent errors (e.g., round-off errors).

- Ensure that the solution algorithm used is the most suitable, and recognize the approximations used.

- Recognize that the accuracy of the solution will only be as good as the accuracy of the input conditions.

- Compare the result with similar problems or simplified versions of the same problem.

4.2   Sensitivity Studies

Most CFD problems depend on mesh quality and resolution. It may be easy to find a suggested mesh density in the literature for a particular problem, but examine the sensitivity of variables such as these on the solution. This may also take an iterative form in which the initial solution has highlighted an area of insufficient mesh resolution or a grid that needs improvement.

Guidelines

- Perform the calculation using several different grid densities.
- Investigate the sensitivity of boundary conditions.
- Run the problem using a different source code and compare the results, if time permits.
- Investigate the effects of different viscous approximations or turbulence models.

4.3   Uncertainties

As described earlier, uncertainties arise through lack of knowledge, which can be a lack of knowledge of the details of the problem to be modeled or of the methods and approximations used to solve the problem. The latter can only be solved by increased user awareness of the theories and methods used. Uncertainties also can occur because of simplification of the problem caused by modeling constraints.

Guidelines

- Avoid oversimplification of the model that may omit important effects.

- Be aware of the magnitude and implication of errors (e.g., round-off errors).

- Recognize the importance of the scale factor. Solution is much easier at model scale (smaller Reynolds number), but there may be difficulties scaling up the results (i.e., Froude and Reynolds scaling differences).

5.0 VERIFICATION AND VALIDATION OF THE CALCULATION AND NUMERICAL MODEL

5.1 <u>Introduction</u>

A code user has responsibility for verification that quantifies and limits discretization error and verification of the initial conditions, boundary conditions, and other special options provided in the input model. However, a code user also shall obtain and review verification documentation from the developers to confirm that the code has been verified adequately. If such documentation is unavailable or inadequate, then appropriate caveats must be provided in documentation of results or the user needs to perform whatever code verification is possible.

To the extent possible, code verification examines implementation of the full mathematical model through comparison to exact analytical results, manufactured solutions (Roache, 2002), or previously verified higher accuracy simulations. Of these options, the method of manufactured solutions is the most powerful, but it requires the user to provide source terms as a function of spatial location and time for every PDE active in the problem solution.

Unfortunately, analytical results and manufactured solutions are only useful for verification of the portions of a code responsible for approximating partial derivatives and solving the system of PDEs associated with the flow problem. They do not help verify coding of complex algebraic expressions used for contributions such as turbulent diffusion coefficients, wall heat transfer functions, reaction rates, and the equation of state. In these cases, rigorous verification is often only possible for code developers. For them, the first step is a rigorous QA procedure. Another best practice for developers is to independently code the algebraic expressions implementing physical models and carefully compare results from the two independent model implementations. At a less rigorous level, developers also should drive implementations of physical models in a separate program and compare directly against the data from which the original algebraic model was obtained. Discrepancies strongly suggest, but do not prove, an error in model implementation.

If the code developer's verification is inadequate, the code user either shall verify the software independently or understand that the validation process may be effectively checking an undocumented model. In this case, extra caution should be taken to validate the model against the full operating space of the system and scenario being simulated. Scaling behavior seen in the documented model is not completely assured in the implemented model.

Comparison to data also can contribute to verification if there is sufficient knowledge about the expected performance of the numerical method or the physical model for a given test case. Usually this information comes from publications or other external sources. This activity provides necessary, but not sufficient, information for comprehensive verification of a code.

Error evaluation for the solution of a particular simulation involves the following subtasks:

- quality assurance of the system input model

- examination of iterative convergence

- basic consistency checks (e.g., checks on global mass, momentum, and energy conservation)

- examination of spatial grid convergence

- examination of temporal convergence

Inconsistencies in any of these checks would be expected to quickly point to implementation problems in the input model (or, on occasion, the software). Once the verification checks have been passed, the validation task can start. The following sections describe some acceptable techniques to perform the verification tasks listed above.

5.2   <u>Error Hierarchy</u>

The range of possible errors in a simulation should be addressed in a logical, hierarchical sequence to obtain efficient error quantification. In the case of CFD software, this sequence starts at round-off errors and then proceeds to iteration errors, discretization errors, and, finally, model errors. The term error hierarchy also implies that numerical errors can be strictly separated from model errors.

Round-off errors are caused by insufficient machine accuracy and can be understood through some relatively simple studies. One quick way to check the effect of round-off errors is to run a relevant simulation with programs compiled with and without optimization engaged. Other variations on this approach include running the same problem on codes generated by different compilers. These techniques do have the disadvantage of being susceptible to undiscovered (but infrequent) compiler errors. Another approach is to execute the simulation with the program adapted for higher precision data and arithmetic than normally used. Most simulations are done with 64-bit representation (double precision) of floating point numbers. If the problem (or relevant portion) can be rerun with a 128-bit representation (quad precision), useful information also will be available on the impact of round-off errors.

Most CFD codes use iterative schemes for matrix solution and for dealing with the coupling and nonlinearities of the underlying equation system. In these cases, insufficient convergence can cause unacceptable errors in final results. Discretization errors should only be investigated when these iteration errors have become sufficiently small.

Discretization errors are the difference between the solution of the discrete approximation to the PDEs in the mathematical model and the actual PDE solution. To obtain mathematically sound solution error estimates, systematic grid size and time-step reduction are necessary. Once the asymptotic range of the convergence properties of the numerical method is reached, the difference between solutions on successively refined grids can be used as an error estimator (Roache, 1998). See Appendix A for more details.

An alternative method to grid or time-step refinement is to analyze the same problem with different discretization schemes followed by a comparison of target variables. The rationale behind this method is that all consistent discretization schemes should converge to the same solution, as the grid is refined toward zero mesh length. Differences in solutions obtained with different discretization schemes on the same grid, point to regions where the grid is still too coarse. Comparison of solutions on the same grid, but with different discretization schemes, is more practical for many 3-D problems than the grid refinement technique discussed above because it does not have the excessive disk space and calculation time requirements (e.g., halving the grid distance in three directions leads to an increase in grid points by a factor of 8 and an increase of calculation times by factors larger than 8). A disadvantage of comparing solutions on the same grid, but with different discretization schemes, is the lack of a mathematical theory for error quantification, as with the Richardson extrapolation (Roache, 1998). Therefore, this method yields only heuristic criteria and indicators. However, it is a fairly economic technique because the cost of going from a first-order scheme to a second-order scheme or from a flux-blending to a second-order scheme is often small. The same arguments apply in an analogous fashion to temporal discretization schemes, if grid size is replaced by time-step.

5.2.1 Target Variables

The first step for solution verification should be the definition of the target variables. Numerical errors should be monitored for a limited number of representative target variables defined during the phenomena Identification and ranking table (PIRT) process (NUREG/CR-6978, 2008) as being representative of the goals of the simulation. It is usually inefficient to evaluate and check all values of all variables. These target variables can, for instance, be maximum or minimum dependent variable values (e.g., maximum or peak cladding temperature, minimum cladding temperature, etc.) or integral quantities, such as mass flow rate and surface heat transfer. Under optimal conditions, these variables are computed during run-time and for steady-state solutions displayed as part of the convergence history. They should be readily available to existing postprocessing tools. Further criteria are:

- sensitivity to numerical treatment and resolution
- computation with existing postprocessing tools
- computation inside the solver and ideally displayed during run-time

The first point is important, since the target variables should be indicative of the numerical errors and uncertainties. The last two points simplify the definition and monitoring of the variables that are especially important for judging iterative convergence.

5.3 Iteration Errors

A first indication of the convergence to the solution is the reduction of the residuals (or residual norms) of the difference equations. However, different types of flows require different levels of residual reduction. For example, swirling flows often can exhibit significant solution changes even when the equation residuals have been reduced by more than 5 to 6 orders of magnitude. Other flows can be well converged with a reduction of only 3 to 4 orders of magnitude. As a result, it also is necessary to monitor the solution during convergence and to plot the pre-defined target quantities of the simulation as a function of the residuals. A visual observation of the solution fields at different levels of convergence is recommended. It is also recommended to monitor the global balances of conserved variables, such as mass, momentum, and energy, during the iterative process. In summary, the following steps can aid the analyst in judging the iterative convergence:

- Plot evolution of residual norms as a function of iteration number.
- Report global mass and heat balance as a function of iteration number.
- Plot target variables as a function of iteration number.
- Report target variables as a function of residual norms.

5.4 Spatial Discretization Errors

Spatial discretization errors result from the use of finite-width grids and the approximation of the differential terms in the model equations by difference operators. Experience shows that only space discretization methods with second- and higher-order truncation errors are able to produce high-quality solutions on practical grids. For some grids, only first-order methods will produce converged steady-state solutions. However, in such cases, solutions must be regarded with caution. The convergence is a result of a numerical viscosity larger than the actual turbulent viscosity. In some instances, the first-order solution can be used successfully as an initial guess at the steady state for a higher-order analysis. In others, the numerical viscosity is simply masking fundamentally transient behavior in the flow.

Since the end user usually cannot change the truncation error order of a given discretization scheme, spatial discretization errors can only be influenced by the provision of optimal grids. It is important for the quality of a solution that the grid points are concentrated in regions of large truncation errors, which are often the regions of large solution variation. It is also important to provide high-quality numerical grids for the reduction of spatial discretization errors.

For mathematically sound grid convergence tests, simulations should be performed on at least three successively refined grids. The target quantities also should be given as a function of the grid size.

Using the Richardson extrapolation (Richardson, 1910), an error estimate in the target variable caused by discretization in space can be made as follows (see Appendix A for details):

$$\varepsilon_1 = \frac{\Theta_1 - \Theta_2}{r^p - 1} \tag{1}$$

In this equation, Θ is the target variable (temperature, velocity, heat-transfer coefficient, maximum temperature, mass flow rate, etc.), r is the grid refinement ratio (always greater than 1), and p is the truncation error order of the discretization scheme. A subscript of 1 indicates results from the finest grid.

An independent estimate of the order of accuracy for the discrete approximation can be obtained from results on three successive grids.

$$p = \frac{\ln\left(\dfrac{\Theta_3 - \Theta_2}{\Theta_2 - \Theta_1}\right)}{\ln(r)} \tag{2}$$

This value of p can only be expected to approach the theoretical accuracy of the numerical method when mesh size is small enough to be in the asymptotic region. In this region, only the lowest-order contribution to truncation error is significant, and the following indicator should be nearly constant:

$$E_h = \frac{\varepsilon}{h^p} \tag{3}$$

50

For practical 3-D simulations, limited computational resources often make it extremely difficult to obtain all three mesh solutions in the asymptotic region. In this situation, useful information on mesh errors can still be obtained by driving subregions of the mesh with appropriate boundary conditions. A code user also should realize that practical implementations of numerical methods (particularly when flux limiters or highly distorted grids are involved) do not always perform at their advertised order of accuracy, even in the asymptotic region. this error estimation procedure does not impose an upper limit on the real error; it is an approximation for evaluating the quality of the numerical results. The best results are obtained for error estimates when:

- Order of accuracy used in Equation (1) is obtained from Equation (2).

- All three meshes are in the asymptotic region.

- The solution at the estimation point is continuous with continuous derivatives.

- Points sampled for error analysis are not too close to inflection points in a plot of the target variable versus the discretized dimension.

- Any iteration employed in the solution is adequately converged.

Even if it is not possible to obtain results for three meshes within the asymptotic range, there is still hope for useful results from a Richardson analysis. The asymptotic range comes from consideration of terms in a classic Taylor series-based truncation error analysis of the discrete approximations to the PDEs; therefore, a Richardson analysis is simply an extrapolation using a curve fit to results from a sequence of mesh refinements in the form:

$$\Theta(h) = \Theta_{exact} + ah^p \tag{4}$$

where h is the relevant mesh (or time-step) size. See Appendix A for clarification of coefficient a in Equation (4.) If results for a target variable at the same spatial location for three grids (Θ_1, Θ_2, and Θ_3) lie on a smooth, monotonic curve, then the use of Equation (1) with Equation (2) can be expected to give a sensible estimate of the error associated with the finest grid. Although the rigor of the results in the asymptotic range is missing, results in this case can still have value in determining regions where a mesh is inadequate.

Roache (1998) deals with quality-of-error estimates through the use of a grid convergence index (GCI) to measure error and a factor of safety (F_s) to cover degradation of the error estimate caused by results from a grid outside the asymptotic range.

$$GCI = F_s \left| \frac{\Theta_1 - \Theta_2}{\Theta_1} \right| \frac{1}{r^p - 1} \tag{5}$$

When the three grids used to calculate p are known to be in the asymptotic region, Roache recommends a value of 1.25 for the factor of safety; otherwise, he recommends a value of 3.0. For unstructured meshes, the above considerations are only valid in case of a global refinement of the mesh. Otherwise, the solution error will not be reduced continuously across the domain. For unstructured grid refinement, the refinement ratio, r, can be defined as follows:

$$r_{effective} = \left(\frac{N_1}{N_2} \right)^{1/D} \tag{6}$$

where N_i is the number of grid points and D is the dimension of the problem. It is also recommended to conduct a visual check on convergence through graphical comparison of selected variables obtained on three grids. The following steps should be followed:

- Define a target variable.

- Provide three (or more) grids using the same topology (or for unstructured meshes, a uniform refinement over all cells).

- Compute solution on these grids and ensure convergence of the target variables.

- Plot selected variables for the different grids.

- Check if the solution is in the asymptotic range.

Plotting experimental data with the results for the two finest grids can provide a quick feel for when it is time to look for other sources of error in the simulation results.

5.5   Time Discretization Errors

To reduce time integration errors for unsteady-state simulations, it is recommended that at least a second-order accurate time discretization scheme be used. For oscillating flows, the relevant frequencies usually can be estimated beforehand, and the time step can be adjusted to provide at least 10 to 20 steps for each period of the highest relevant frequency. In case of unsteadiness caused by a moving front, the time-step should be selected as a fraction of:

$$\Delta t \approx \frac{\Delta x}{U}$$

In this equation, Δx is the grid spacing and U is the front speed. Sometimes, under strong grid and time-step refinement, flow features can be resolved that are not relevant for the simulation.

In principle, the time dependency of the solution can be treated as another dimension of the solution error estimation. However, a 4-D grid study would be very demanding. Therefore, it is more practical to perform the error estimation in the time domain separately from the space discretization. Starting with a sufficiently fine space discretization, the error estimation in the time domain can be performed as a 1-D study.

Studies should be performed with at least two and, if possible, three different time steps for one given spatial resolution. The error estimators given in Section 5.4 can be used if the spatial grid size is replaced by the time-step. The following information should be provided:

- unsteady-state target variables as function of time step (graphical representation)
- error estimate based on (time-averaged) target variables
- comparison with experimental data for different time-step values

5.6   Quality Assurance

The most important step in error control is to understand that errors will occur regardless of the method used to generate source code or input models. Procedures shall be in place to eliminate (or at least minimize) programming or user input errors. QA procedures are a proven way to control the introduction of bugs and formalize test procedures. These procedures work well for both code development and application input model development.

Four key components of QA are documentation, development procedures, testing, and review. Written standards for these components should be established at the beginning of a project and accepted by everyone involved. Documentation of a new code or new simulation usually begins with a simple statement of requirements for what must be modeled, what approximations are and are not acceptable, and the form of implementation. A complete written description of the underlying mathematical model provides a basis for verification activities. A clear description should be provided of relevant experiments for use in validation activities. Any uncertainties in the input model and in code models also should be described for later studies of sensitivity of results to model uncertainties. A test plan describes calculations based on the validation experiments and any necessary verification tests, including discretization error studies described in previous sections.

Documentation should be generated in two drafts. The first precedes the creation of the model. When used for a specific system simulation, this initial QA document builds on documentation from the PIRT process. The second draft is issued as a final report and includes the final form implemented and the results of all proposed sensitivity tests. Both drafts should be accompanied by two phases of independent review. The first focuses on the viability of the proposed approach, and the second focuses on the completeness of testing and sensitivity studies. Combined review of the documentation and the CFD model is a powerful technique for catching and correcting errors. Even before an independent review, the act of describing implementation with words forces a careful consideration of the CFD model.

CFD users also should have a reference QA document providing procedures for input model creation that reduce chances for initial introduction of errors without significant reduction in the developer's effectiveness.

Code users also need guidelines for documentation of input models and, where appropriate, style and order of creation of the input model. However, the first significant step in controlling user errors is to include the quality of the user interface(s) as a strong component in the selection criteria for a CFD software package. Look for capabilities that minimize opportunity for simple typographic errors and provide clear, easily accessible guidance in option selection. Important features to consider include:

- the ability to define geometry directly from CAD files
- automated aids to mesh generation
- menu-driven interface for option selection and specification of initial conditions
- direct links to useful documentation describing each menu and menu item
- error checking on inconsistent input and values that are out of range

For QA of CFD user input, consideration should be given to a more automated form of source file documentation through a configuration management procedure. This starts with a systematic record of all changes, dates of change, and individuals responsible for the changes.

5.7   Validation of Results

Once the verification process has limited discretization and iteration convergence errors to acceptable levels, validation of physical models can proceed. This section discusses basic considerations for validation, as well as the associated uncertainty analysis needed to build final validation metrics and to confirm completeness of the validation.

5.7.1   Validation Methodology

In the field of CFD, the real world is modeled first by a conceptual model (governing equations), and then by a computational model (computer code). Application of the computer code or, more specifically, of one concrete computational path to a scientific or industrial problem, leads to a computational solution. The computational solution should be validated.

Validation is a process of determining the degree to which a model is an accurate representation of the real world from the perspective of the intended uses of the model (AIAA Guide, 1998). Here, the "real world" is a system (engineering hardware) for which a reliable engineering simulation tool is needed. Such a system typically is very complex with many coupled physical phenomena taking place in complicated geometry. Therefore, a tiered approach is recommended for validation of models of such systems. In Oberkampf and Trucano (2002) and Oberkampf et al. (2004), the following four progressively simpler tiers are defined:

- complete system
- subsystem cases
- benchmark cases
- unit problems

Careful attention to the tiered approach minimizes one of the most insidious problems in code validation: cancellation of errors. Confidence is built in relevant models contributing to the CFD simulation by first testing isolated physical processes and simple geometries and then moving up through testing with higher levels of complexity in process interaction.

Validation of a CFD code should then start from the unit problems, in which only one element of complex physics is allowed to occur in each problem so that elements of complex physics are isolated as tractable items. Unit problems are characterized by very simple geometries, often 2-D or 3-D, with important geometric symmetry features. Experiments should be on highly instrumented test facilities producing highly accurate data supported by extensive uncertainty analysis of the data for validation calculations at this level. If possible, repeated runs should be performed, even at separate facilities, to aid in identifying random and systematic (bias) errors in the experimental data. All of the important code input data, initial conditions, and boundary conditions should be accurately measured and documented. If some significant parameters needed for the CFD simulation were not measured, reasonable or plausible values for the missing data should be assumed. In this case, an estimation of the possible effect of missing information on computed results should be performed. A rigorous (and seldom feasible in the CFD field) approach in this case requires multiple computations and a statistical uncertainty analysis to estimate sensitivity of target variables to the possible range of unknown (or uncertain) system parameters.

Benchmark cases typically involve only two or three types of coupled flow physics in more complex geometry than in the unit problems, retaining the features critical to these types of physics. Most of the required modeling data, initial conditions, and boundary conditions are measured, but some of the less important experimental data are not measured in some cases. As in the case of unit problems, whenever assumed values are used to replace missing input data, uncertainty analysis should be performed. For subsystems and complete systems, it is difficult and sometimes impossible to quantify most of the test conditions required for thorough validation of the computational model. Three or more types of physics are coupled (some coupling reduction is typical for subsystem cases). Some of the necessary or the most important modeling data, initial conditions, and boundary conditions are measured. Typically, there is less experimental data and less measurement precision provided at this level than in the case of unit problems and benchmark cases. Taken as standalone validation, these factors reduce reliability of detailed conclusions on suitability of the computational model to the intended application. However, taken in conjunction with unit and benchmark tests, subsystem and complete system tests provide necessary validation of interactions between individual process models. Traditional experiments are intended to improve understanding of the physical world and have been classified by Oberkampf et al.(2004) as follows:

- Physical-discovery experiments conducted primarily for the purpose of improving the fundamental understanding of some physical process.

- Experiments conducted primarily for construction or improving mathematical models of fairly well-understood flows.

- Experiments that determine or improve the reliability, performance, or safety of components, subsystems, or complete systems.

Validation experiments have the primary goal of quantifying differences between a portion of the physical world and the equivalent portion of a virtual world. As a result, design of a validation experiment requires both skilled experimentalists and individuals with detailed knowledge of the contents and behavior of the simulation tool (both developers and code users). The experiment should be designed to answer questions about a specific application, and the PIRT process should guide the design to capture the essential physics of interest and to measure state variables most sensitive to the relevant model implementations in the code. Special care should be taken with the experiment to obtain initial and boundary conditions for use in the simulation. This includes precision measurements of hardware geometry and instrument location rather than the use of dimensions from design drawings. These data, as well as data from instrumentation during the experiment, should be accompanied by reliable estimates of random (precision) and bias (systematic) errors. In the case of initial and boundary conditions, these errors form the basis of uncertainty analysis for key computed results. For physical state data, these errors should be included in consideration of validation metrics.

Scoping studies with the simulation code may provide guidance to the design of the experiment. However, the communication of results between experimentalists and analysts should end when it is time to actually perform the experiment and simulation of the experiment. Results from the two groups should be obtained independently and only compared after each activity is completed. It is common to perform a second post-test round of simulation, but care should be taken to ensure that changes to the input model only reflect differences in initial and boundary conditions between design and actual execution of the experiment.

The last step in the validation process is formulation of conclusions. Validation cannot be understood as a binary ("yes" or "no") problem. From an engineering viewpoint, validation is an estimation problem: What is the measure of agreement between the computational result and the experimental result, and how much is the measure affected by numerical error in the computational solution and by the experimental uncertainty? The answers are clearly application- and user-dependent. Acceptance criteria are in most cases determined very vaguely, and there is also a risk of faulty conclusions. There is a "model builder's risk" (i.e., risk of rejection of a model when the model is actually valid based on errors on both the computational and the experimental side), and there is also a "model user's risk" in accepting the validity of a model when the model is actually invalid and the original favorable agreement has compensating, or cancelling, errors in the comparisons. Oberkampf and Trucano (2002) state that compensating error in complex simulations is a common phenomenon. It is also well known that model the user's risk is potentially higher because it produces a false sense of security. It is also more likely to occur in practice because of the tendency to find agreement of results and not to spend more time and resources pursuing possible problems in either the computations or the experiments.

5.7.2 Target Variables and Metrics

A panel of experts should select target variables for validation during the PIRT process. Because PIRT is recognized to be an iterative process, the list of target variables may change as experience is gained with the experiment or with computational scoping studies. Note that target variables may be fundamental quantities, such as velocity, temperature, and pressure, or derived quantities, such as flow rates, heat-transfer coefficients, or a maximum, minimum, or average over more fundamental data.

Selection of suitable validation metrics is an important part of the validation process. Oberkampf and Barone (2006) provide a detailed discussion of considerations for selecting metrics. Two key considerations are that the metric include a comparison to a reliable measure of experimental uncertainty, and that the presentation of the metric values does not include qualitative judgments, such as "very good agreement."

To obtain reliable values for experimental uncertainty, results should be available from redundant validation experiments. With the data from multiple runs of the same experiment, a basic metric would be the difference between a computed value and the mean of the experimental values at the same location, presented with a confidence interval for the experimental data. In this case, the metric involves statement of three numbers: (1) the estimated error between results of the simulation and the true experimental value, (2) an estimated range within which the true value of this error lies, and (3) the confidence level that the value lies within the quoted range (usually selected as 90 percent or 95 percent for the statistical analysis). Useful global metrics can be constructed by integrating the local error estimates or corresponding fractional errors over time or space, as appropriate. However, the corresponding integration of confidence intervals (or intervals ratio to the mean experimental value) simply become confidence indicators because of loss of rigor in the interpretation of the resulting interval. Care also shall be taken in using such global metrics because regions with relatively large error may be masked by the averaging process.

To place the metric in the proper perspective, information on experimental error also should be provided that, to the extent possible, clearly distinguishes between truly random error and systematic (bias) error. Consider the hypothetical comparison in Figure 5-1 of calculated and measured mass flow rate at a specific location. The error bars could be the result of phenomena that vary randomly with time during any run of the experiment. Another possibility is that they reflect calibration error resulting in a fixed offset (bias) of data in any given experiment. This offset might vary randomly from experiment to experiment as a result of the calibration process. In later evaluation of validation metrics, the nature of the experimental error in Figure 5-1 can make a significant difference in conclusions about the quality of one or more models used in the numerical simulation. If the error is truly random within each experiment, it might be concluded that the simulation adequately captures the physical phenomena. However, if the error is a bias, the simulation misses a key trend in the data and, depending on the needs for the final application, one or more relevant models could be judged to be inadequate.

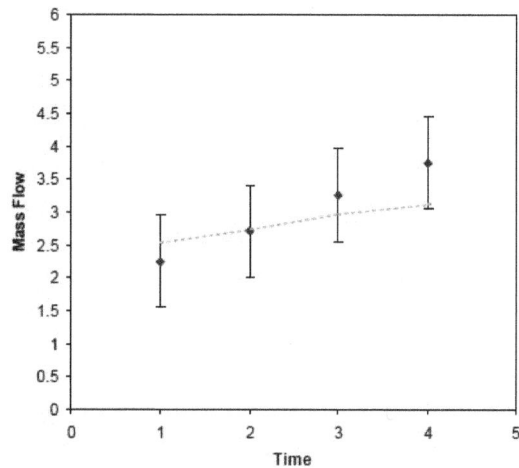

Figure 5-1 Comparison of a calculation to data

57

6.0   CHECKLIST FOR CFD BEST PRACTICE

This section contains CFD best practice guideline (BPG) checklists. The lists roughly follow the chronological sequence required to complete a CFD analysis. As applicable, the guideline checklists also can be used to review CFD analysis prepared by a cask vendor to support the thermal evaluation.

6.1   Guidelines on the Training of CFD Users

- CFD users for nonroutine applications should have good training and knowledge in classical fluid mechanics, a broad understanding of numerical methods, and detailed knowledge of the application being examined. This means that they will be able to understand the limitations of the particular models used (e.g., turbulence, boundary conditions).

- The training and education requirement for more routine applications can be less stringent, provided that clear guidelines or procedures have been established for the use of the code. An example of a routine application would be the simulation of convection flow in a ventilated storage cask (VSC), which has been performed for many previous designs; and for which only relatively small changes in geometry and boundary conditions are expected to occur.

- In both routine and nonroutine applications, training on the use of the specific CFD code with the solution of realistic exercises is needed.

6.2   Guidelines on Problem Definition

The analyst should carefully think about the requirements and objectives of the analysis and may consider the following points:

- Is a CFD simulation appropriate for the problem in question?

- Are the objectives of the analysis clearly defined?

- What are the requirements for accuracy? (This would depend on the target value. For example, in typical calculations an energy and mass balance error of about 5 percent would be acceptable, while the error in PCT would be expected to be within the GCI value discussed later in this guide.)

- What local or global quantities are needed from the simulation?

- What are the documentation and reporting guidelines, as indicated in the Standard Review Plans?

- What are the important flow physics involved?

- What is the area of primary interest (domain) for the flow calculation?

- Is the geometry well defined?

59

- What level of validation is necessary?

- Is the applicant providing an analysis approach for which validation has been previously performed on similar applications, or is this a new approach for which adequate validation has not been completed?

- What level of computational resources is needed for confirmatory analysis (memory, disk space, CPU time) and is this available to the analyst?

- Are the details of the analysis model sufficient to capture all important components and effects?

- Are the magnitude and implication of any identified errors reasonable (e.g., round-off errors)?

6.3   Guidelines on Definition of Geometry

- Check and document that the geometry of the modeled component represent the geometry as intended. For example, the transfer of geometrical data from a CAD system to a CFD system may involve loss of surface representation accuracy. Reviewing the geometry as displayed in the CFD analysis code may be helpful.

- In general, it is not necessary to explicitly include geometrical features that have dimensions below that of the local grid size, provided they are taken into account in the modeling (e.g., wall roughness).

- In areas where local detail is needed, grid refinement in local areas with fine details should be used, such as in the area of fine edges or small clearance gaps. If grid refinement is used, the additional grid points should lie on the original geometry and not simply be a linear interpolation of more grid points on the coarse grid.

- Check that geometry is defined in the correct coordinate system and with the correct units for the CFD code. CAD systems often define the geometry in millimeters, and this must be converted to SI units if the code assumes that the geometry information is in these units. Most codes commonly do this.

6.4   Guidelines on Grids and Grid Design

- Clean up CAD geometry, and for body-fitted grids check that the surface grid conforms to the CAD geometry.

- Avoid highly skewed cells in particular. For hexahedral cells or prisms, the included angles between the grid lines should be optimized so that the angles are about 90 degrees. Angles with less than 40 degrees or more than 140 degrees often result in a deterioration in the results or lead to numerical instabilities, especially in the case of transient simulations.

- The angle between the grid lines and the boundary of the computational domain (the wall or the inlet- and outlet-boundaries) should be close to 90 degrees. This recommendation is stronger than the recommendation for the angles in the flow field far away from the domain boundaries.

- Avoid the use of tetrahedral elements in boundary layers.

- Ensure that the aspect ratio (the ratio of the sides of the elements) away from boundaries is not too large. Typically, this aspect ratio should not be larger than 20. This restriction may be relaxed near walls; in fact, it can be beneficial to resolve the boundary layer and the laminar sublayer.

- Observe the code requirements (usually documented in the code user's guide) of mesh stretching or expansion ratios (rates of change of cell size for adjacent cells). The change in mesh spacing should be continuous and mesh size discontinuities should be avoided, particularly in regions of high gradients.

- Ensure that the mesh is finer in critical regions with high-flow gradients, such as regions with high shear, and where there are significant changes in geometry. Make use of local refinement of the mesh in these regions in accordance with the selected turbulence wall modeling. The location of a refinement interface should be away from high-flow gradients.

- Check the analysis results for regions that exhibit high-flow gradients and, if necessary, rearrange grid points.

- Analyze the suitability of the mesh by a grid-dependency study (this could be local) in which at least three different grid resolutions are used (see Appendix A).

6.5 Guidelines on Solution Strategy

Having established a clear problem definition, the user needs to translate this into a solution strategy involving issues such as:

- mathematical and physical models
- pressure- or density-based (coupled) solution method
- turbulence model
- available code/solver
- computational mesh
- boundary conditions

6.6 Guidelines on Turbulence Modeling

- The user should be aware that there is no universally valid general model of turbulence that is accurate for all classes of flows. Validation and calibration of the turbulence model is necessary for all applications.

- If possible, the user should examine the effect and sensitivity of results to the turbulence model by changing the turbulence model being used.

- The relevance of turbulence modeling only becomes significant in CFD simulations when other sources of error (in particular, the numerical and convergence errors) have been removed or properly controlled. Clearly, no proper evaluation of the merits of different turbulence models can be made unless the discretization error of the numerical algorithm is known and grid sensitivity studies have been completed.

61

Guidelines on Standard Wall Functions

- The meshing should be arranged so that the values of y^+ at all the wall adjacent mesh points are between 30 and 100 (The form usually assumed for the wall function is not valid much for y^+ smaller than 30). Some commercial CFD codes account for low values of y+ by switching to alternative functions if y^+ is less than 30. Be aware of this and check the user manuals.

- Cell-centered schemes have their integration points at different locations in a mesh cell than they do in cell vertex schemes. Thus, the y^+ value associated with a wall-adjacent cell differs according to which scheme is used on the mesh. Care should be exercised when calculating the flow using different wall function schemes on the same mesh.

- The values of y^+ at the wall-adjacent cells strongly influence the prediction of friction and, hence, drag. Thus, particular care should be given to the placement of near-wall meshing if these are important elements of the solution.

- Check that the correct form of the wall function is being used to take into account the wall roughness. (Most CFD codes set the wall roughness to zero, by default.) Check the wall-function model for inclusion of roughness to ensure it is not over specified.

Guidelines on Near-Wall Resolution

- Make sure that the turbulence model used is capable of resolving the flow structure through to the wall.

- Ensure that the value of y^+ at the first node adjacent to the wall is close to unity.

- Use a small stretching factor for progressing the mesh spacing away from the wall. There should be at least 10 mesh points between the wall and y^+ equal to 20.

Guidelines on Weaknesses of the Standard k-ε Model

- The turbulent kinetic energy (k) is overpredicted in regions of flow impingement and reattachment leading to poor prediction of the development of flow. Ince and Launder (1995) have proposed a modification to the transport equation for the turbulence dissipation rate (ε) that is designed to tackle this problem.

- Regions of recirculation in a swirling flow are underestimated. Reynolds stress models (RSMs) should be used to overcome this problem.

- Highly swirling flows generally are poorly predicted because of the complex strain fields. RSMs or nonlinear eddy viscosity models should be used in these cases.

- Mixing is poorly predicted in flows with strong buoyancy effects or high streamline curvature. RSMs or a low Reynolds number turbulent model should be used in these cases.

- Flow separation from surfaces under the action of adverse pressure gradients is poorly predicted. The real flow is likely to be much closer to separation (or more separated) than the calculations suggest. The Baldwin-Lomax (1978) one-equation model is often better than the standard k-ε model in this respect. The SST version of Menter's k-ω based, near-wall resolved model also offers a considerable improvement.

- Flow recovery following re-attachment is poorly predicted. Avoid the use of standard wall functions in these regions for this type of problem.

- The spreading rates of wakes and round jets are predicted incorrectly. The use of nonlinear k-ε models should be investigated for these problems. In the nonlinear k-ε model, higher-order terms of the turbulence stress-strain relationship are used (e.g., Speziale nonlinear k-ε, Suga quadratic k-ε model, etc. (Speziale, 1987, 1988; Thangam, 1991).

- Turbulence-driven secondary flows in straight ducts of noncircular cross section are not predicted at all. Linear eddy viscosity models cannot capture this feature. Use RSM or nonlinear eddy viscosity modeling.

- Laminar and transitional regions of flow cannot be modeled with the standard k-ε model. This is an active area of research in turbulence modeling. No simple advice can be given. Potential options include user intervention to switch the turbulence model on or off at predetermined locations. Another option is to use low Reynolds k-ε, as well as the transitional SST version of Menter's k-ω based.

Guidelines on Boundary Conditions

General Guidelines on Boundary Conditions

- Ensure that appropriate boundary conditions are available for the case under consideration. For swirling flows, consult the applicable CFD user's guide to ensure that an appropriate boundary condition is used (e.g., radial equilibrium of pressure field instead of constant static pressure). Special nonreflecting boundary conditions are sometimes required for outflow and inflow boundaries where strong pressure gradients exist.

- Check if the CFD code allows inflow at open boundary conditions. If inflow cannot be avoided at an open boundary, ensure that the transported properties of the incoming fluid (including turbulence boundary conditions) are modeled properly.

- Examine the possibilities of moving the domain boundaries to a position where the boundary conditions are more readily identified, well posed, and can be precisely specified.

- Perform a sensitivity study for each application in which the boundary conditions are systematically changed within certain limits to see the variation in results. If any of these variations prove to have a sensitive effect on the simulated results, more accurate data on the specified boundary conditions should be obtained.

Guidelines on Inlet Conditions

- Examine the possibilities of moving the domain inlet boundaries to a position where the boundary conditions are easily identified, well posed, and can be precisely specified.

- For each application, a sensitivity analysis should be performed in which the inlet boundary conditions are systematically changed within certain limits. Aspects that should be examined are:

  o inlet flow direction and magnitude
  o uniform inlet velocity (slug flow) or velocity profile
  o variation of physical parameters
  o variation of turbulence properties at inlet

Guidelines on Specification of Turbulence Quantities at an Inlet

- A particularly important issue is the specification of the turbulence properties at the inlet to the computational domain. Verified quantities should be used as inlet boundary conditions for turbulent kinetic energy k and dissipation ε, if these are available, since their magnitude can significantly influence the results.

- If no data are available, the values shall be specified using sensible engineering assumptions and the influence of the choice should be examined with sensitivity studies.

- For the specification of the turbulent kinetic energy k, values appropriate to the application should be used. These values generally are specified through a turbulence intensity level. ERCOFTAC guidelines suggest a variety of values depending on flow type. In hydrodynamics, low "inlet" turbulence levels are likely, but zero turbulence will bring about anomalies in turbulence modeling unless specialized approaches to laminar and transitional regions are adopted.

- The specification of the turbulent length scale, as an equivalent parameter for the dissipation ε, is more difficult. For external flows, a value determined from the assumption that the ratio of turbulent and molecular viscosity μ_t/μ of the order of 10 is appropriate. For simulations in which the near-wall region is modeled, the length scale should be based on the distance to the wall and be consistent with the internal modeling in the code.

- If more sophisticated distributions of k and ε are used, they shall be consistent with the velocity profile so that the production and dissipation term in the turbulence equations are in balance. An inconsistent formulation such as a constant velocity profile and constant profile of turbulence intensity at the inlet lead to an immediate unrealistic reduction of the turbulence quantities after the inlet. These can be checked by making a plot of the ratio of turbulent to molecular viscosity μ_t/μ. In cases where problems arise, the inflow boundary should be moved sufficiently far away from the region of interest so that an inlet boundary layer can develop.

- For RSM models, the stresses need to be specified and, as these normally are not available, an assumption of isotropic flow conditions with zero shear stresses is generally made.

Guidelines on Outlet Conditions

- The boundary conditions imposed at the outlet should be selected to have a weak influence on the upstream flow. Extreme care shall be taken when specifying flow

64

velocities and directions on the outlet plane. The most suitable outflow conditions are weak formulations involving specification of static pressure at the outlet plane.

- Particular care should be taken in swirling flows in which the pressure distribution on the outlet boundary is strongly influenced by the swirl and cannot be specified independently of the swirl coming from upstream.

- Be aware of the possibility of inlet flow inadvertently occurring at the outflow boundary, which may lead to difficulties in obtaining a stable solution or even to an incorrect solution. If it is not possible to avoid this by relocating the position of the outlet boundary in the domain, one way to potentially avoid this problem is to restrict the flow area at the outlet, provided that the outflow boundary is not near the region of interest.

- If there are multiple outlets, impose either pressure boundary conditions or mass flow specifications, depending on the known quantities.

Guidelines on Solid Walls

- Boundary conditions imposed on solid walls should be consistent with both the physical and numerical models used (e.g., convection to and from wall, radiation to and from wall, wall thickness and conductive resistance, roughness).

- If roughness on the wall is not negligible, significant levels of uncertainty can arise through incorrect specification of roughness within the wall function.

Guidelines on Symmetry and Periodicity Planes

- Symmetry and periodicity planes assume that the gradients perpendicular to the plane are either zero (for symmetry) or determined from the flow field (periodicity). If symmetry or periodicity planes cross the inlet or outlet boundaries, inlet or outlet variables should be specified to be consistent with them.

Guidelines on Uncertainties with Steady Flow, Symmetry, and Periodicity

- Check carefully if the geometry is symmetric or if a geometrical distortion or disturbance in the inlet conditions is present that can trigger asymmetric solutions. Estimate the Reynolds number of the inflow and check if the flow could be asymmetric, turbulent, or unsteady (Roache, 1994, 1998).

- After obtaining a steady solution, switch to the transient mode and check if the solution remains stable.

- If there are difficulties getting a converged steady solution, especially if there is an oscillation of the residuals, switch to the transient mode.

- In case of doubt, the simulation should be unsteady and without symmetry assumptions as boundary conditions.

Guidelines on Global Solution Algorithm

- Check the adequacy of the solution procedure (what algorithm is used in terms of interpolation scheme to calculate flow field variables (e.g., semi-implicit method for pressure-linked equations (SIMPLE), SIMPLE-revised, etc.) for the physical properties of the flow.

- As a first step in this process, the parameters controlling convergence (e.g., relaxation parameters or Courant number) of the solution algorithm should be used, as suggested by the CFD-code vendor or developer.

- If it is necessary to change parameters to aid convergence, it is not advisable to change too many parameters in one step, as it then becomes difficult to analyze which of the changes influenced the convergence. In the case of persistent divergence, see the sections in this document on boundary conditions, grid, discretization, and convergence errors.

- If a steady solution has been computed and there is a reason to be unsure that the flow is really steady, then an unsteady simulation should be performed with the existing steady flow field as the initial condition. Examination of the time development of the physical quantities in the locations of interest will identify if the flow is steady or not.

6.7 Guidelines on the Solution of the Discretized Equations

Guidelines on Round-Off Errors

- Always use the 64-bit representation of real numbers (double precision on common UNIX workstations).

- Developers are recommended to use the 64-bit representation of real numbers (REAL*8 in FORTRAN) as the default settings for their CFD code.

Guidelines on Spatial Discretization

- Give an approximation of the numerical error in the simulation by applying a mesh refinement study or by mesh coarsening (see Appendix A).

- If available in the code, make use of the calculation of an error estimator (which may be based on residuals or on the difference between two solutions of different order accuracy).

Guidelines on Temporal Discretization

- The overall solution accuracy is determined by the lower-order component of the discretization. At least second-order accuracy is recommended in space and time. For time-dependent flows, the time and space discretization errors are strongly coupled. Therefore, finer grids or higher-order schemes are required (in both space and time).

- Check the influence of the time step on the results.

- Ensure that the time step is adapted to the choice of the grid and the requested temporal size by resolving the frequency of the realistic flow, and ensure that it complies with eventual stability requirements (heat transfer in spent fuel transfer cask neutron shield).

6.8   Guidelines on Assessment of Errors

- A potential source of user errors is in the implementation of the solution strategy with a particular code. Such errors might be minimized by using a formal checklist or letting another CFD analyst check the code input data.

- Questions that should be considered include the following:

  - Have the boundary conditions been properly defined and properly applied?

  - Has the appropriate system of units been used?

  - Is the geometry correct?

  - Are the correct physical properties specified?

  - Have the intended physical and mathematical models been used (e.g., gravity forces, radiation, user defined subroutines, etc.)?

  - Have default parameters that may affect the solution been changed?

  - Have the appropriate convergence criteria been defined and used?

6.9   Guidelines on Analysis and Interpretation of Results

Guidelines on Checking Results

- Ensure that an elementary interpretation of the flow-field explains the fluid behavior and that the trends of the flow analysis can be reconciled with a simple view of the flow.

- Check conserved variables, including an overall mass and heat balance.

- Check that temperatures, velocities, pressures, etc., have realistic values.

- Check if fluid variables, such as velocity and pressure, are smoothly distributed over the body and vary rapidly only where expected. Discontinuities may be the result of a poor or insufficient mesh.

- Perform simple hand calculations to check orders of magnitudes of variables.

- Run simple versions of the problem (e.g., with reduced geometry) to get an idea for the numbers involved.

- Ensure that the mean values of engineering parameters derived from the simulation are computed consistently (e.g., mass-averaged values, area-averaged values, time-averaged values). Calculation of local and mean engineering parameters with external postprocessing software may be inconsistent with the solution method of the code used. Check that any test data used for comparison with the simulations also is computed in the same way as the data from the simulation.

Guidelines on the Relevance of the Results

- Consider whether the interpretation of the results and any decisions made are within the accuracy of your computation.

- Ensure that the solution algorithm used is the most suitable. Be aware of the approximations used in the algorithm.

- The accuracy of the solution will only be as good as the accuracy of the input parameters.

- Compare the result with similar problems, or simplified versions of the same problem.

Guidelines on Further Sensitivity Studies

- Perform the calculation using several different grid densities.

- Investigate the sensitivity of boundary conditions.

- Run the problem using a different source code and compare the results, if time permits.

- Investigate the effects of different viscous approximations or turbulence models.

6.10 Guidelines on Documentation

- Keep accurate records of the simulation with clear documentation of assumptions, approximations, simplifications, geometry, and data sources.

- Organize the documentation of the calculations so that another CFD expert can follow what has been done.

- Be aware that the level of documentation required depends strongly on the applicant's requirements as defined in the problem definition.

6.11 Guidelines for the Code User

- The user should recognize that codes can only be validated and verified for a class of problems involving specific variables. If the user is moving into an area in which the code is not fully verified, there is a greater risk of code errors.

- A suite of test cases set up and run by the user on new code releases provides an independent check on the code and highlights changes between releases (e.g., in default parameters).

- When a code error is suspected, the user should communicate this to the code vendor or developer as soon as possible, especially if no list of known bugs has been published. Other users may then profit from this experience, or the user may find that the bug is well known and a solution or workaround is available.

- In communication with the code developer or code vendor about a suspected program error, the user should provide a concise description of the problem and all the necessary input data files so that the error can be reproduced. In cases in which commercial sensitivity precludes this, special arrangements will need to be made.

6.12 Guidelines on Convergence

- Be aware that different codes have different definitions of residuals.

- Always check the convergence on global balances (conservation of mass, momentum, and turbulent kinetic energy) where possible, such as the mass flow balance at inlet and outlet and at intermediate planes within the flow domain.

- Check not only the residual itself but also the rate of change of the residual with increasing iteration count.

- Convergence of a simulation should not be assessed purely in terms of the achievement of a particular level of residual error. Carefully define solution-sensitive target quantities for the integrated global parameters of interest and select an acceptable level of convergence based on their rate of change (such as mass flow, wall heat transfer, temperatures, and velocities).

- For each application, perform a test of the effect of converging to different levels of residual on the integrated parameter of interest. (This can be a single calculation stopped and restarted at different residual levels.) This test demonstrates at what level of residual the parameter of interest can be considered to have converged and identifies the level of residual that should be aimed for in similar simulations of this class of problem.

- Monitor the solution in at least one point in a sensitive area to see if the region has reached convergence.

- For calculations proving difficult to converge, the following advice may be helpful:

  o Use more robust numerical schemes during the first (transient) period of convergence and switch to more accurate numerical schemes as the convergence improves.

  o Reduce parameters controlling convergence (e.g., under relaxation parameters).

  o If the solution is heavily under-relaxed, increase relaxation factors at the end to see if the solution holds.

  o Check if switching from a steady to a time-accurate calculation has any effect.

  o Consider using a different initial condition for the calculation.

  o Check the numerical and physical suitability of boundary conditions.

o   Check if the grid quality in areas with large residual has any effect on the convergence rate.

o   Look at the residual distribution and associated flow field for possible hints (e.g., regions with large residuals or unrealistic velocity levels).

7.0   APPLICATION

The following application example is provided to illustrate some of the issues described in these guidelines. The example illustrates the modeling routines that CFD users practiced in the application of CFD to dry cask analysis. Data for a ventilated storage cask (VSC-17) collected by Idaho National Laboratory (INL) was used to validate a 3-D CFD model. In this example, validation was used to reduce the modeling and application uncertainties. To address the modeling uncertainties, the example study focused on turbulence modeling of buoyancy-driven air flow. Similarly, in the application uncertainties, the pressure boundary used to model the air inlet and outlet vents was investigated and validated. The VSC-17 experiment consisted of 17 assemblies loaded with consolidated pressurized-water reactor (PWR) spent fuel. At the time of the tests, the fuel was generating about 14.9 kilowatts (kW) of decay heat. The experiments were performed with vacuum, nitrogen, and helium backfill cavity environments in a vertical orientation . Solid block was used to model the fuel rods. As a result, no flow or convection heat transfer was considered. This assumption was acceptable in this case because consolidated fuel rods were loaded in the assemblies. The effective thermal conductivity model k_{eff} (TRW report, 1996) was used to model the combined effect of thermal radiation and conduction within the assembly.

7.1   Validation

7.1.1   Purpose of the Analysis

The objective of this task was to validate a general purpose CFD method to perform thermal evaluations of the VSC-17 system. In addition, the effectiveness and validity of the effective thermal conductivity model k_{eff} (TRW report, 1996) was quantified and verified. The k_{eff} model is used to represent the combination of radiation and conduction heat transfer by an equivalent thermal conductivity in the region housing the spent fuel. The present analysis used FLUENT, a commercially available CFD software package. FLUENT is a finite control volume-based code suited to model all modes of heat transfer and fluid flow in open-flow regions of a storage system. As such, there is a need to investigate the applicability of the k_{eff} model in CFD codes such as FLUENT, which is capable of modeling all modes of heat transfer. Among the tests performed at INL (as shown in S-2 (McKinnon et al., 1992), test #1 was selected to be modeled with FLUENT.

Two types of flows exist in spent fuel dry storage casks such as VSC-17. Inside the sealed canister, compressed helium flows through the fuel rod assemblies because of buoyancy forces, while air flows outside the canister in an open system manner as a result of buoyancy (density difference). The standard k-ε model with standard wall function often is used to bridge the viscous layer near the wall to the fully turbulent core region in the middle of a flow channel. Therefore, the second objective of this validation is to compare the performance of different turbulence models and laminar flow.

7.1.2   VSC-17 Spent Fuel Storage Cask Experiments

The performance tests were conducted on a Pacific Sierra Nuclear VSC-17, a ventilated concrete storage cask containing consolidated PWR spent fuel. The Pacific Northwest Laboratory and INL performed the experimental work.

The VSC-17 spent fuel storage system is a passive heat dissipation system for storing 17 assemblies or canisters of consolidated spent nuclear fuel. The VSC-17 system consists of a ventilated concrete cask (VCC) enclosing a multi-assembly sealed basket (MSB) containing spent nuclear fuel as shown in Figure 7-1. Decay heat generated by the spent fuel is transmitted through the containment wall of the MSB to a cooling air flow. Natural circulation drives the cooling air flow through an annular path between the MSB and the VCC liner and carries the heat to the environment without undue heating of the concrete cask. The annular air flow cools the outside of the MSB and the inside of the VCC.

The cask weighs about 72,575 kilograms (80 tons) empty and 99,790 kilograms (110 tons) loaded with 17 canisters of consolidated fuel. The VCC has a reinforced concrete body with an inner steel liner and a weather cover (lid). The MSB contains a guide sleeve assembly for fuel support and a composite shield lid that seals the stored fuel inside the MSB. Different tests were undertaken as shown in Table 7-2. From these tests, test #1 was selected to validate the FLUENT model. In this test, the cavity atmosphere in the MSB is helium at slightly subatmospheric pressure. The helium atmosphere inside the MSB enhances the overall heat transfer capability and prevents oxidation of the fuel and corrosion of the basket components.

The performance tests consisted of loading the MSB with 17 fuel cans containing consolidated PWR spent fuel from Virginia Power's Surry reactors and Florida Power and Light's Turkey Point reactors. Based on pretest predictions, fuel rod decay heat generation rates totaled about 14.9 kW during testing. Cask surface, concrete, air channel surfaces, and fuel canister guide tube temperatures were measured. Testing was performed with the cask in a vertical orientation and with vacuum, nitrogen, and helium backfill environments in the MSB. During the tests, air circulation vents were open, partially blocked, and completely blocked, as shown in Table 7-2. Test #1 (fully open vents) with helium gas in the MSB was selected for validation. Additional details of the experiment can be found in the original documentation of the testing (McKinnon et al., 1992).

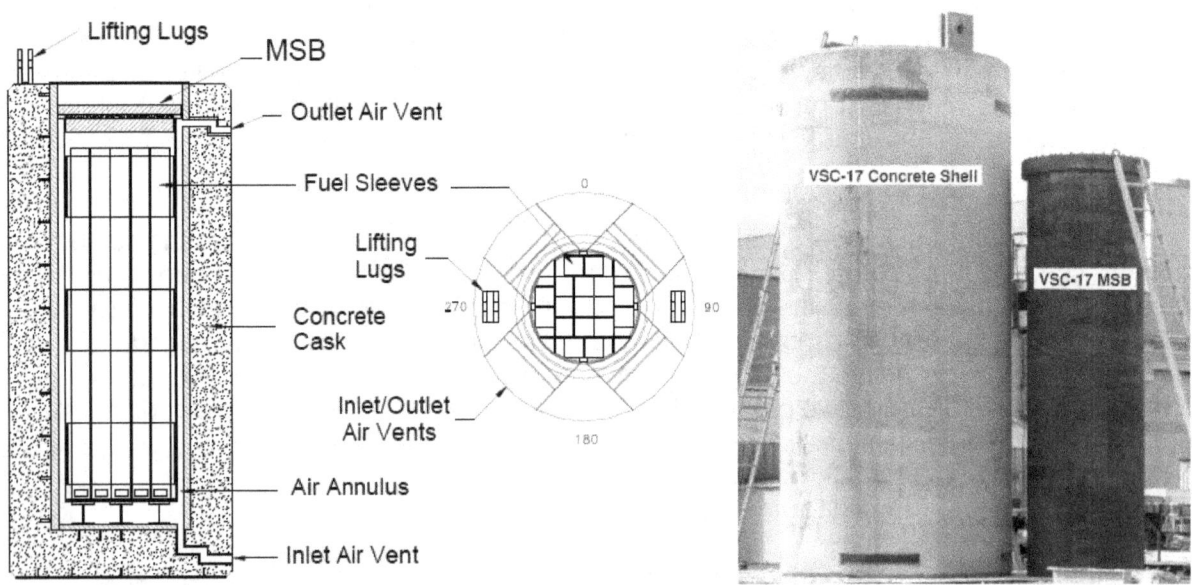

Figure 7-1 Ventilated concrete cask system

7.1.3 Effective Thermal Conductivity Model for Consolidated Fuel Canister

Spent nuclear fuel generates heat as a result of continuing radioactive decay. All heat generated within the fuel rods must be rejected to the fuel basket. Heat is transported from the fuel rods to the fuel basket structure by the parallel action of conduction and thermal radiation heat transfer. Fuel assemblies possess considerable mass and, therefore, have substantial thermal capacity. The fuel thermal capacity affects the response of a dry cask system to transient thermal loadings.

It is impractical and computationally very expensive to model every fuel rod in every stored fuel assembly explicitly. Instead, the cross section bounded by the inside of the storage cell, which surrounds the assemblage of fuel rods and the interstitial helium gas, is replaced with an equivalent square characterized by an effective thermal conductivity. The effective thermal conductivity of the cell space will be a function of temperature because radiation heat transfer (a major component of heat transport between the fuel rods and the assembly walls) is a strong function of the temperatures of the participating bodies. Therefore, each storage cell location will have a different value of effective conductivity (depending on the coincident temperature) in the homogenized model.

The thermal modeling properly recognizes the assembly as an anisotropic conduction media. In particular, the inplane conductivity and the axial conductivity of the fuel region differ considerably. This is because, while inplane heat transfer is interrupted by gaps, in the axial direction heat flows in an uninterrupted manner in the fuel rods. The thermal property of the 3-D assembly continuum is characterized by both planar and axial conductivities.

The first step in these analyses is to construct and evaluate a geometrically accurate planar model of a fuel assembly using the FLUENT code. This model includes full details of the fuel rods, including fuel pellets, the rod fill gas, and the cladding. The interstitial space between the fuel rods and assembly walls is occupied by the backfill gas and, therefore, has the thermal conductivity of the gas. A uniform volumetric heat generation is applied to the fuel pellet regions, and uniform temperature is applied to the model periphery (the inner periphery of the fuel storage cell). Both conduction and thermal radiation heat-transfer mechanisms are included and steady-state temperature distribution is obtained for a range of edge temperatures.

A decay heat of 0.5 kW to 1.2 kW was considered to cover the entire decay heat range generated by each assembly stored in VSC-17. Wall temperatures ranging from 93 degrees Celsius (C) to 400 degrees C were assumed to obtain the inplane effective thermal conductivity. Once the temperature gradient is obtained, the temperature-dependent fuel effective thermal conductivity is determined using the k_{eff} methodology described in the TRW report (1996). Negligible differences were observed for the inplane effective thermal conductivity for the different heat rates.

The effective axial thermal conductivity was determined using an area-weighted fraction of the material in the fuel cell space. This relationship was implemented in FLUENT based on the temperature-dependent thermal conductivity of cladding material. As a result of the long, uninterrupted axial heat transfer paths in the fuel assemblies, the effective axial thermal conductivity will be greater than the corresponding planar conductivity, which is limited by the numerous gaps in the heat flow path.

The radial and axial k_{eff} values for helium environment are provided in Tables 7-7 and 7-8. The model is implemented in FLUENT as temperature-dependent k_{eff} values. FLUENT's option for anisotropic thermal conductivity was used to represent the different effective conductivities of the fuel region in the axial and radial directions.

7.1.4 Decay Heat Generation for Consolidated Fuel Cans

A total canister decay heat of 14.9 kW was predicted near the start of the testing for an average of 877 watts (W) per canister. Decay heat from individual canisters ranged from 700 W to 1,050 W per consolidated canister. Canister placement was selected to create ⅛ symmetry of heat generation within the basket and to produce a maximum fuel temperature in the center of the MSB. Because of the imposed decay heat symmetry, thermocouples were used to measure temperature values only in one quadrant, as shown in Figure 7-3. The peripheral assemblies in this quadrant have decay heat values of about 744 W. The assemblies next to the peripherals have decay heat values close to 970 W, while the central assembly has a maximum decay heat of 1,050 W.

The heat generation rates for these fuel cans were applied uniformly to the homogeneous volumes representing the fuel and backfill gas inside each assembly of the FLUENT model. An axial power profile (based on the measured axial power distribution shown in Figure 7-2a) was applied to each assembly.

7.1.5 Mesh Consideration and Turbulence Modeling in Air Annulus Region

The selection of a turbulence model for the air flow through the annular gap between the VCC and MSB is directly tied to the type of mesh density in that region. The mesh size near the wall depends on the selected turbulence model. The near-wall modeling significantly affects the fidelity of numerical solutions, inasmuch as walls are the main source of mean vorticity and turbulence. It is in the near-wall region that the solution variables have large gradients, and the momentum and other scalar transports occur most vigorously. Therefore, accurate representation of the flow in the near-wall region determines successful predictions of wall-bounded turbulent flows. In this study, two types of mesh distributions were used in the annular region. The first mesh was designed to allow the use of "standard wall functions" to bridge the viscosity-affected region between the walls and the fully turbulent core region. The use of wall functions obviates the need to modify the turbulence models to account for the presence of the wall. This type of modeling usually is applicable to high Reynolds number flows. The second mesh was designed to allow resolution of the viscosity-affected region all the way to the wall, including the viscous sublayer. This type of approach is referred to as the "near-wall modeling" approach. The predicted dimensionless distance between the wall and the cell center near the wall (y^+) for the first mesh was around 20, whereas for the second mesh it was around 1.

To determine the flow regime, both Reynolds and Grashof numbers were calculated in the air channel between the MSB and the VCC, as well as in the helium region inside the MSB. In the air-flow region, a Reynolds number above 3,000 was predicted based on the channel hydraulic diameter and air maximum velocity in the annulus. This is clearly above the critical Reynolds number of 2,300 for internal flows. This flow is clearly in the transitional range between the laminar and turbulent regimes. Since these are buoyancy-driven flows, both the Grashof number based on the hydraulic diameter of the channel and the modified Grashof (Gr) number, defined as ($Gr_m = Gr*W/H$) where W and H are the width and height of the air channel, were also calculated. Based on the Grashof number and modified Grashof number, laminar flow was predicted, as shown in Sparrow and Azevedo (1985). On the other hand, buoyancy-driven helium flow inside the canister was predicted to be in laminar regime, based on both Grashof

74

and Reynolds numbers. Because of a higher kinematic viscosity and the low achieved velocities of the inert gas within the MSB, a low Reynolds number of approximately 200 was calculated. This is clearly in the laminar flow regime.

As mentioned earlier, two different grids were used to represent the air annular and inlet and outlet regions to use and compare different types of turbulence models. The standard k-ε turbulence model with standard wall functions was used to model turbulence for the coarser mesh. Both the transitional SST k-ω and the low Reynolds k-ε were used to model turbulence. Near-wall modeling (mesh resolution, integrated all the way to the wall) was used to model turbulence for the finer mesh. In addition, since the calculated Reynolds number was close to the critical Reynolds number of 2,300, a laminar model with finer mesh was also considered.

7.1.6 Thermal Radiation Modeling within the VSC-17 System

Radiation heat transfer was represented by the discrete ordinate (DO) model. In this approach, the radiative transfer equation (RTE) for an absorbing, emitting, and scattering medium is solved for a finite number of discrete solid angles. The fineness of the angular discretization is controlled by the user. Unlike the discrete transfer radiation model (DTRM), the DO model does not perform ray tracing. Instead, the DO model transforms the RTE equation into a transport equation for the radiation intensity in the spatial coordinates (x, y, z). The DO model solves for as many transport equations as there are directions defined by the angular discretization. The solution method is identical to that used for the fluid flow and energy conservation equations. For the VSC-17 analysis, four angular discretizations were used in each direction of the spherical coordinates system (theta (θ) and phi (φ)).

7.1.7 Boundary Conditions

Two types of control volumes were used. The first control volume, "extended control volume," included the dry cask and a portion from the surrounding environment. Pressure was applied at the boundaries outside the dry cask, and a convection heat-transfer coefficient was used at the top lid of the cask. The second control volume included the dry cask only and used free heat-transfer coefficient at the top lid and vertical surface. These coefficients were based on free convection correlation in still air for vertical and horizontal surfaces. At the bottom surface of the cask, equivalent resistance to heat conduction to underlying soil was used to prescribe an equivalent heat transfer coefficient. Since the measurements were performed indoor, solar insolation was not included, but radiation between surfaces inside the cask and from the cask outer surfaces to the ambient was included.

The following boundary conditions are applied to the FLUENT VSC-17 model:

- ambient temperature of 21 degrees C (measured)

- ambient pressure applied at the external boundaries for the extended control volume

- ambient pressure applied at the inlet and outlet vents for dry cask

- heat-transfer coefficient of 5 W/m^2-K on the top and side of the VCC

- heat-transfer coefficient of 10 W/m^2-K on the top of the VCC weather cover

- heat-transfer coefficient of 0.17 W/m^2-K on the bottom surface of the VCC (equivalent to conduction through 3 meters (m) of soil)

- soil temperature of 15 degrees C

Depending on the surface material, different emissivity values were used. The most complete set of data are listed in Sucec (1985) and McAdams (1954). These textbooks reference an emissivity of 0.4 for fuel cans (made of 304 stainless steel), 0.6 for basket, basket supports, and MSB body (made of A516 steel), and 0.7 for A36 steel with some oxidation used for VCC annulus and inlet and outlet liners.

The consolidated fuel cans were modeled as nonporous solid using the effective thermal conductivity values obtained from the 2-D FLUENT thermal model of a single assembly, as shown in Figures 7-8a and 7-8b.

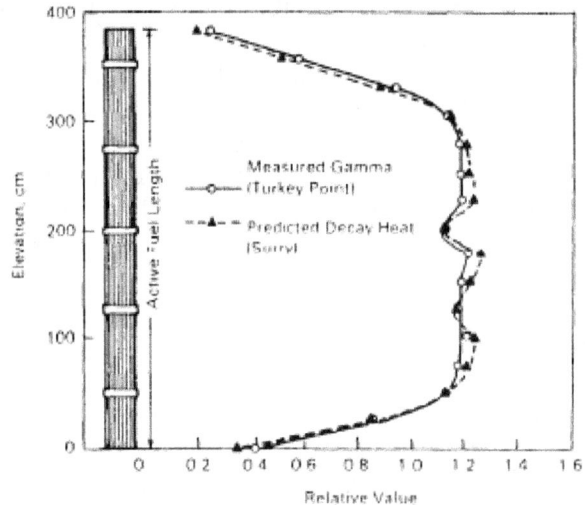

Figure 7-2a Decay heat axial profile

7.1.8 Material Properties

The VSC-17 solid material thermal properties are tabulated in the test documentation (specifically, Table 5.2 in McKinnon et al. (1992)). Temperature-dependent thermophysical properties and ideal gas law were used for the working fluids (air and helium.) The fluid thermophysical properties were obtained from JANAF (1985).

Table 7-1 Material Thermophysical Properties (McKinnon et al., 1992)

Material	Thermal Conductivity W/m-°C (Btu/ft-hr-°F)	Density Kg/m³	Heat capacity J/kg-K
Air	Kinetic theory	Ideal Gas Law	Cp(T) (JANAF, 1985)
Helium	Kinetic theory	Ideal Gas Law	Cp(T) (JANAF, 1985)
Concrete	1.47 (.85)	n/a	n/a
Steel liner (A36)	41.5 (24)	n/a	n/a
Steel basket assembly (A512)	41.5 (24)	n/a	n/a
Steel fuel cans (SS 304)	16.3 (9.4)	n/a	n/a
RX-277 (radiation shield in lid)	0.52 (0.3)	n/a	n/a

7.1.9 Spatial Differencing and Solution Method

FLUENT's SIMPLE method was used to solve the steady-state equations that govern the flow and heat transfer in the VSC-17 model. Second-order upwind spatial differencing was used for all variables except the pressure equation (continuity equation), in which a body force weighting factor was used.

These simulations were run from a zero-flow initial condition using pressure boundary at the airflow inlet. The criterion for solution convergence is typically when the total heat flux is within 20 W, corresponding to an energy error of approximately 0.5 percent.

7.1.10 Cases Considered

Extended Control Volume

The extended control volume included the cask and a portion of the surrounding environment (Figure 7-6). Two grid sizes of 1.2 million cells and 2 million cells were considered to check for a grid independent solution. In the finer (2 million cells) mesh case, the additional cells were placed near the surrounding VCC and MSB walls, the air flow duct, and walls surrounding inert gas flow regions inside the canister. Only the transitional SST k-ω model was used in this part of the analysis to model the turbulence in the air flow region. As pressures are used at the inlet and outlet boundaries, an input for operating density is required. Therefore, in this part of the analysis, several cases were considered to investigate the effect of operating density on the predicted results. Most of the performed analyses presented in this report are based on the dry cask model only, and the extended model is used to determine the operating density. For additional information about the operating density, see Fluent (2006).

Also, in this part, the measured inlet temperature was checked for consistency. When reviewing the experimental data, it was noticed that the average temperature at the inlet was at least 13 degrees C higher than the ambient temperature for test #1. This step was of utmost importance because of the major difference that can be obtained in the final results when different values of the inlet temperature are used in the analysis. Several cases were run to investigate the influence of the ambient temperature at the inlet of the dry cask model. These cases are shown in the results and discussion section of this document. The effect of using different turbulence models was investigated using the dry cask model. Among the turbulence models available in FLUENT, transitional k-ω SST, low Reynolds k-ε, and standard k-ε models were considered to model the air flow. The laminar model was also considered because of the low Grashof and modified Grashof numbers in the air channel gap, as explained in Sparrow and Azevedo (1985).

7.1.11 Thermal Performance Data

Ninety-eight thermocouples (TCs) were used to measure the thermal performance of the cask. The inside of the MSB was instrumented with seven TC lances, as shown in Figure 7-2b. Each TC lance contained six calibrated Type J (iron-constantan) insulated junction TCs that provided a total of 42 internal lance TCs. Fifty-three Type J TCs were used to determine the temperature of the MSB, cask lid, and concrete. Ten TCs were attached to the outer surface of the cask, five were attached to the MSB lid, two were attached to the weather cover, 10 were imbedded in the concrete, nine were attached to the outside barrel of the MSB, nine were attached to the inner liner of the VCC, and one was installed in the center of each air inlet and outlet vent. Three additional TCs were used to monitor the ambient temperature in the Hot Shop. The location of the TC lances and the elevations of the TCs are shown in Figure 7-3.

Each TC lance had six TCs installed in an 8-mm-diameter (0.315-inch) tube, as shown in Figure 7-2b. Lances were inserted through instrumentation penetrations in the test lid and into selected guide tubes placed in six fuel canisters and into one simulated guide tube attached to the MSB. The selected axial and cross-sectional locations of the lance TCs made it possible to evaluate temperature symmetry and to determine axial and radial temperature profiles for the cask.

Table 7-2 identifies the performance test runs and provides conditions associated with each test, including backfill gas, vent blockage, and internal MSB pressure, along with the test run number and date. The steady-state temperatures measured in test #1 are provided in McKinnon et al. (1992). The location of each measurement can be determined from TC information in Figures 7-2b and 7-3.

Dimensions in mm

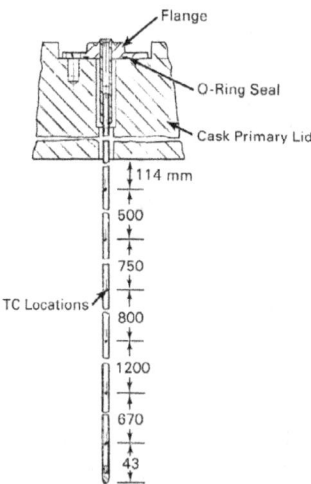

Figure 7-2b Thermocouple lance

78

Dimensions in mm

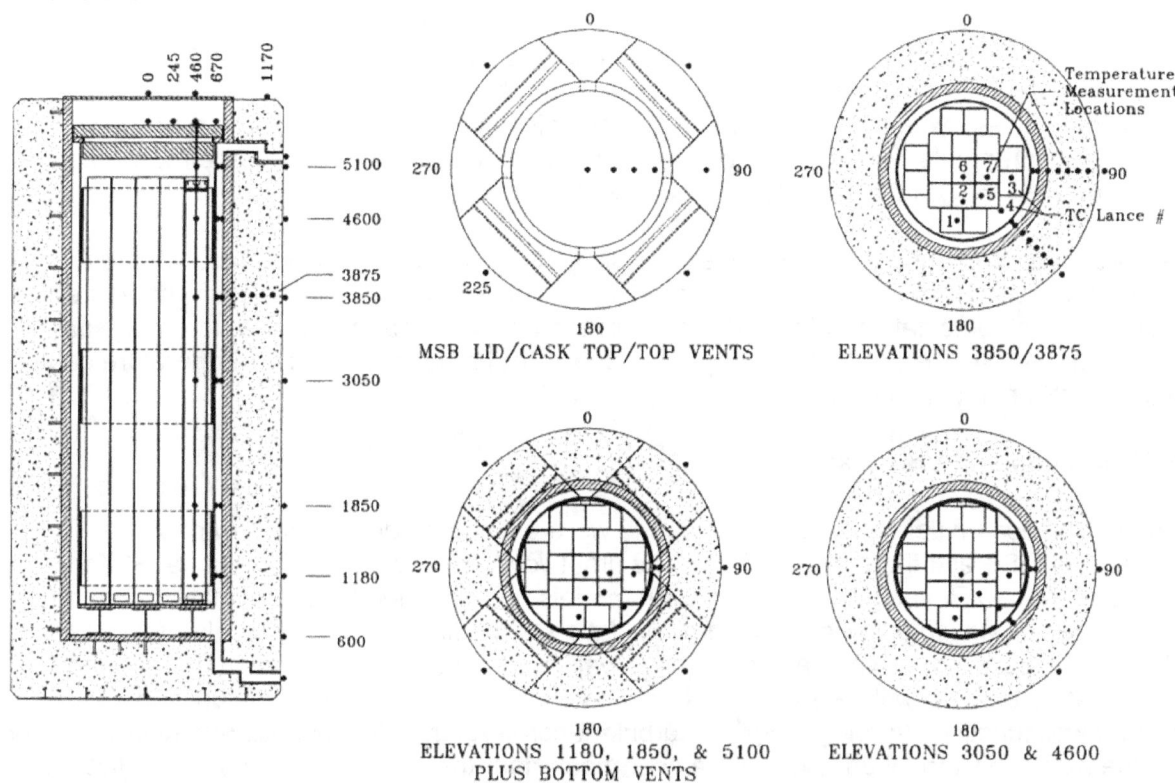

MSB LID/CASK TOP/TOP VENTS

ELEVATIONS 3850/3875

ELEVATIONS 1180, 1850, & 5100
PLUS BOTTOM VENTS

ELEVATIONS 3050 & 4600

Figure 7-3 Temperature measurement locations used during the VSC-17 performance test

Table 7-2 Performance Test Run Designation

Test #	1	2	3	4	5	6
Backfill gas	Helium	Helium	Helium	Helium	Nitrogen	Nitrogen/Vacuum
Pressure, mbar absolute	817.5	1,074.1	935.3	975.2	843.6	8.6
Vent blockage (P-1/2 Inlets, I-inlets, A-all inlets and outlets)	None	P-Block	I-Block	A-Block	None	None
1990 test dates	Dec. 11	Oct. 23	Nov. 26	Nov. 29	Nov. 13	Nov. 15

7.1.12 Measurement Uncertainty

Temperature uncertainties for the internal thermocouples are within $\pm$4 degrees C, and external temperature measurements are within $\pm$4.5 degrees C, based on the combined uncertainties of the thermocouples, extension wires, and data acquisition system. The higher accuracy of the internal measurements was obtained because the thermocouples were calibrated whereas the thermocouples used on the cask outside surface were not.

Pressure measurement uncertainties were within $\pm$1.5 mbar for the low-pressure vacuum measurement and within $\pm$6 mbar for the reading near 1,500 mbar. The pressure measurement uncertainty is a combination of the uncertainty of the transducer and the precision of the data acquisition system. Detailed uncertainty calculations for both pressure and temperature can be found in McKinnon et al. (1992).

7.1.13 Results and Discussions

Three turbulence models and a laminar regime were used to model the air-flow passage between the MSB and the concrete liner. The first two models were the transitional SST k-ω model and the low Reynolds k-ε model. Both models use damping functions that consider the effect of the cell Reynolds number on the calculation of the time and length scale of turbulence. Both models are used with a fine grid near the wall ($y^+{\sim}1$) to enable integration through the viscosity-affected near-wall region. The third model is the standard k-ε in conjunction with standard wall functions to bridge the fully turbulent core region to the viscosity-dominated region near the wall. This model does not use a finer mesh near the wall. In the present application, a y^+ close to 20 was used.

Calculated temperature profiles using the four approaches described above are compared to the experimental data. Measured axial temperature profile for Lances 3, 5, 6, and 7 inside the fuel region, liner wall, and MSB wall were selected to compare to the calculated CFD results. In addition, measured radial profiles from the center of the fuel region to the periphery of the overpack concrete shield at elevations 3.05 m and 3.85 m were used to compare to the CFD predicted results.

Initially, the extended control volume case (shown in Figure 7-6) was used. The modeled volume included the dry cask and a portion of the surrounding ambient. Two types of grids were used to check for a grid independent solution. Only the transitional SST k-ω model was used in this part of the analysis to model the turbulence in the air-flow region. Figures 7-9a through 7-9c show the CFD temperature distributions along with measured data. The first grid consisted of 1.2 million cells. Figures 7-10a through 7-10c show the analysis results along with measured data for the second grid, which consisted of 2 million cells. The additional cells in the second grid were packed near the walls, where noticeable velocity and temperature gradients are expected. The figures indicated that both grids resulted in the same temperature distribution for the MSB wall, liner wall, and Lances 3, 5, 6, and 7. The similarity is also reflected in the radial distribution at elevations 3.05 m and 3.85 m. Both grids indicated the same PCT as shown by Lance 6 in Figures 7-9a and 7-10a.

As stated earlier, the extended control volume case also was used to find the appropriate operating density for the dry cask model (shown in Figures 7-4, 7-5, 7-7a, and 7-7b.) The FLUENT user manual (Fluent, 2006) indicates that the operating density should be equivalent to the volumetric average density of the fluid. In the extended control volume, the air volume consists of the extensive ambient air and the air sandwiched between the liner and the MSB walls. Since the volume of the hot air between liner and MSB walls is negligible compared to the represented ambient volume, the average density is closer to the ambient density and so is the operating density. For the dry cask model, the air volume consists of only the air between the VCC liner and the MSB walls. This air volume continuously removes heat from the MSB wall, which increases the air temperature from inlet to outlet. The average fluid temperature of this volume is higher than the ambient temperature and so is the operating density (Fluent, 2006). The inlet boundary condition to the dry cask model is equal to the average measured ambient temperature.

The results for this simulation are shown in Figures 7-11a through 7-11c. The results were identical to the results obtained for the extended control volume, indicating that the problem was correctly modeled (i.e., correct operating density). The operating density used for this case corresponded to the inlet temperature and not the volumetric average as suggested by the FLUENT user manual (Fluent, 2006). When the same case was run with the operating density as the average fluid average density (suggested by the FLUENT user manual), the air mass flow entering VSC-17 was 50 percent smaller and the air temperature at the outlet was 12 degrees higher, as shown in Table 7-6. Table 7-6 also indicates that less heat was absorbed from the fuel rods, which resulted in a higher PCT. Figures 7-12a and 7-12b show the difference in axial distribution for Lance 6, MSB wall, and liner walls when different operating densities were used for the dry cask model. When a lower operating density is used, these figures show that temperatures inside the MSB and the air channel are consistently higher. If a value for the operating density is not provided in the CFD model, FLUENT will use the average density of all modeled fluids. For this case, it will use the volumetric average density of air and helium. As helium density is much lower than air density for the given conditions, the volumetric average density will be even lower than the air volumetric average density used previously. Therefore, a lower air mass flow rate and a higher PCT will be obtained for an average operating density based on air and helium.

The measured inlet temperatures shown in McKinnon et al. (1992) for test #1 were about 12 degrees higher than the measured ambient temperature. There is no explanation for the temperature at the inlet vents to be higher than the ambient temperature. McKinnon et al. (1992) indicates an inlet average temperature of 35 degrees C for test #1, yet the measured cask surface temperature just above the inlet is 27 degrees C for test #1. Either the thermocouples at the inlet vents were reporting wrong measurements or they were affected by radiation from inside the cask. Regardless of the reason, these thermocouples most likely were not measuring the correct inlet temperature. To reconcile this issue, the extended control volume CFD case, which does not need the inlet temperature to be provided since the inlet vents are part of the domain, is used to predict the inlet vent temperature for the dry cask model. FLUENT results show that the temperature at the entry of the dry cask is identical to the ambient temperature as shown in Table 7-3. Table 7-4 shows the exit temperature for the dry cask model using three different turbulence models for test #1. These cases used the average measured inlet temperature as the inlet temperature boundary. The turbulence models used are the low Reynolds k-ε, the transitional k-ω SST, and the standard k-ε. As shown in Table 7-4, all three turbulence models led to the same higher temperature. The resulting exit temperature is 10 degrees higher than the average exit temperature shown in McKinnon et al. (1992).

Table 7-5 shows the exit temperature for the dry cask model using three different turbulence models. These cases used the ambient temperature as an inlet temperature boundary condition. The turbulence models used are the low Reynolds k-ε, the transitional k-ω SST, and the standard k-ε. As shown in Table 7-5, all three turbulence models predicted very well the exit temperature for the dry cask model when compared to measured data. This is yet further proof and indication that the measured inlet temperature values do not correctly reflect the ambient inlet temperature. As a result, the average ambient temperature was used as the inlet temperature boundary for the dry cask model.

Figures 7-11a through 7-11c, 7-13a through 7-13c, 7-14a through 7-14c, and 7-15a through 7-15c show the temperature distributions for the different turbulence models and laminar flow, based on the dry cask model. The figures show the axial distribution of the fuel inside the canister, MSB, and liner walls and the radial distribution from the center of MSB to the outside surface of the VSC-17 at elevations of 3.05 m and 5.85 m. As a first observation, all four options used to model the turbulence in the air cooling channel successfully predicted the location of the PCT. The PCT value is of great importance in dry cask analysis. For long-term normal storage conditions, PCT is limited to 400 degrees C to avoid fuel cladding deformation caused by excessive creep and to limit the amount of radially oriented hydrides (NUREG-1536, 2010).

Both transitional k-ω SST and low Reynolds k-ε turbulence models (applied in the air annular gap) predicted the temperature distribution fairly well in the fuel region inside the canister, as well as the passage of cooling air (i.e., MSB and liner walls). Both Figures 7-11a and 7-13a show that these two models predicted the location and value of the PCT. In addition, the temperature axial profile of the liner wall and MSB wall were fairly well predicted given the complex nature of this buoyancy-driven flow as shown in Figures 7-11b and 7-13b. The improvement in the prediction of the liner wall distribution was the result of the fine mesh used near the walls and the capability of these two models to handle low Reynolds turbulent flow. In addition, the radial temperature distributions at 3.05 m and 3.85 m compare favorably using these two models as shown in Figures 7-11c and 7-13c.

Table 7-5 shows that when the standard k-ε model is used, a slightly higher air mass flow rate is induced, thus higher heat rate is absorbed from the cask, resulting in a lower air exit temperature. Table 7-5 shows that when the laminar regime is used, a lower air mass flow rate is induced. Consequently, lower heat is absorbed from the cask and higher PCT is predicted.

CFD results obtained for the laminar option are shown in Figures 7-15a through 7-15c. The laminar model overpredicted the PCT and the axial temperature distribution in the entire fuel region as shown in Figure 7-15a. The liner and the MSB wall axial temperature distributions were consistently higher than the experimental data when laminar regime was used as shown in Figure 7-15b. The overprediction of the temperature distribution inside the cask and the air channel led to the overprediction of the radial temperature profile in the overpack region as shown in Figure 7-15c.

The standard k-ε model was a better choice than the laminar option, but because of the lack of a finer mesh near the MSB wall and the liner wall, the model was unable to capture the exact temperature distribution at the liner wall. The model overpredicted the heat exchange between the two walls. Usually, the standard k-ε model combined with standard wall functions is applicable to high Reynolds number flows. This model is not suitable for transitional buoyancy-driven flows.

The standard k-ε model underpredicted slightly the MSB wall distribution and poorly predicted the liner wall distribution. As shown in Table 7-5, the standard k-ε model overpredicted the heat transfer from MSB wall to the liner, which causes a lower temperature on the liner wall. The reason behind this outcome is that the standard k-ε is not well suited for low Reynolds

82

numbers and transitional flows. Integration to the wall (i.e., finer mesh near the wall) is required to get more information about the wall shear stress and heat transfer for low Reynolds number and transitional flows. In transitional Reynolds number flows (as in the VSC-17 case), some type of damping function in conjunction with a fine mesh near the wall is required to enable computation across the laminar viscous sublayer.

It has been shown that low Reynolds k-ε and transitional k-ω SST turbulence models performed better than the standard k-ε for the VSC-17 analysis case. In the standard k-ε model, the viscosity-affected inner region is not resolved. Instead, semi-empirical formulas called "wall functions" are used to bridge the viscosity-affected region between the wall and the fully-turbulent region. The use of wall functions obviates the need to modify the turbulence models to account for the presence of the wall. Low Reynolds k-ε and transitional k-ω SST required a finer mesh near the wall to resolve the viscosity-affected region of the boundary layer. Damping functions dependent on the local Reynolds number are used in the conservation equation of the turbulent kinetic energy and its dissipation to reflect the wall presence and viscous damping. In this approach, the turbulence models are modified to enable the viscosity-affected region to be resolved with a mesh all the way to the wall, including the viscous sublayer. This approach is usually termed the "near-wall modeling" approach. In most high Reynolds number flows, the wall function approach substantially saves computational resources because the viscosity-affected near-wall region, in which the solution variables change most rapidly, does not need to be resolved. The wall function approach is popular because it is economical, robust, and reasonably accurate. It is a practical option for the near-wall treatments for industrial flow simulations. The wall function approach, however, is inadequate in situations in which the low Reynolds number effects are pervasive (as in the case of the dry cask) and the hypotheses underlying the wall functions cease to be valid. Such situations require near-wall models that are valid in the viscosity-affected region and accordingly integrable all the way to the wall.

The low Reynolds number k-ε model is preferred to the transitional k-ω model because gravity effect is included in the production of turbulence and dissipation (Fluent, 2006).

To see the effect of using higher inlet temperature at the boundaries, several cases were considered using the measured average temperature. These cases are shown in Figures 7-16a through 7-16c, Figures 7-17a through 7-17c, and Figures 7-18a through 7-18c. These figures show the axial distribution in the fuel region, MSB, and liner wall, and the radial distribution at elevations 3.05 m and 3.85 m. Each set of figures was obtained using transitional k-ω SST, low Reynolds k-ε, and standard k-ε models, respectively. As seen in these figures, the temperature distributions were overpredicted using both transitional k-ω SST and low Reynolds models. The standard k-ε model compared better with the experimental data when a higher temperature was used at the inlet boundary condition. This, of course, does not mean that the standard k-ε is a good model for this type of flow, since the analysis is based on an unrealistic higher inlet temperature.

Figure 7-19 shows the contour temperature on the weather cover on the top of the dry cask using the low Reynolds turbulence model. The scale in this figure is refined to correctly identify the range of temperature existing on top of the cask. This figure shows that the temperature on the weather cover is almost constant and has a value of around 52 degrees C at a radius of 460 mm (i.e., two-thirds away from the center). The weather cover temperature measured is 52 degrees C at the center and 53 degrees C at a radius of 460 mm, as reported in McKinnon et al. (1992). The figure shows that the heat transfer, including conduction, convection, and radiation, was analyzed correctly in the axial direction.

Figure 7-20 shows the outside surface axial temperature distribution based on the low Reynolds k-ε turbulence model. The CFD analysis followed the trend and correctly predicted the measured temperatures. This figure shows how well the heat transfer and fluid flow was modeled radially. Figure 7-21 further illustrates these temperature contours on the outside surface. Figure 7-22 shows the temperature contours for both air and helium flow using the low Reynolds k-ε turbulence model. This figure shows that the PCT occurs towards the cask middle because of the slightly subatmospheric canister pressure.

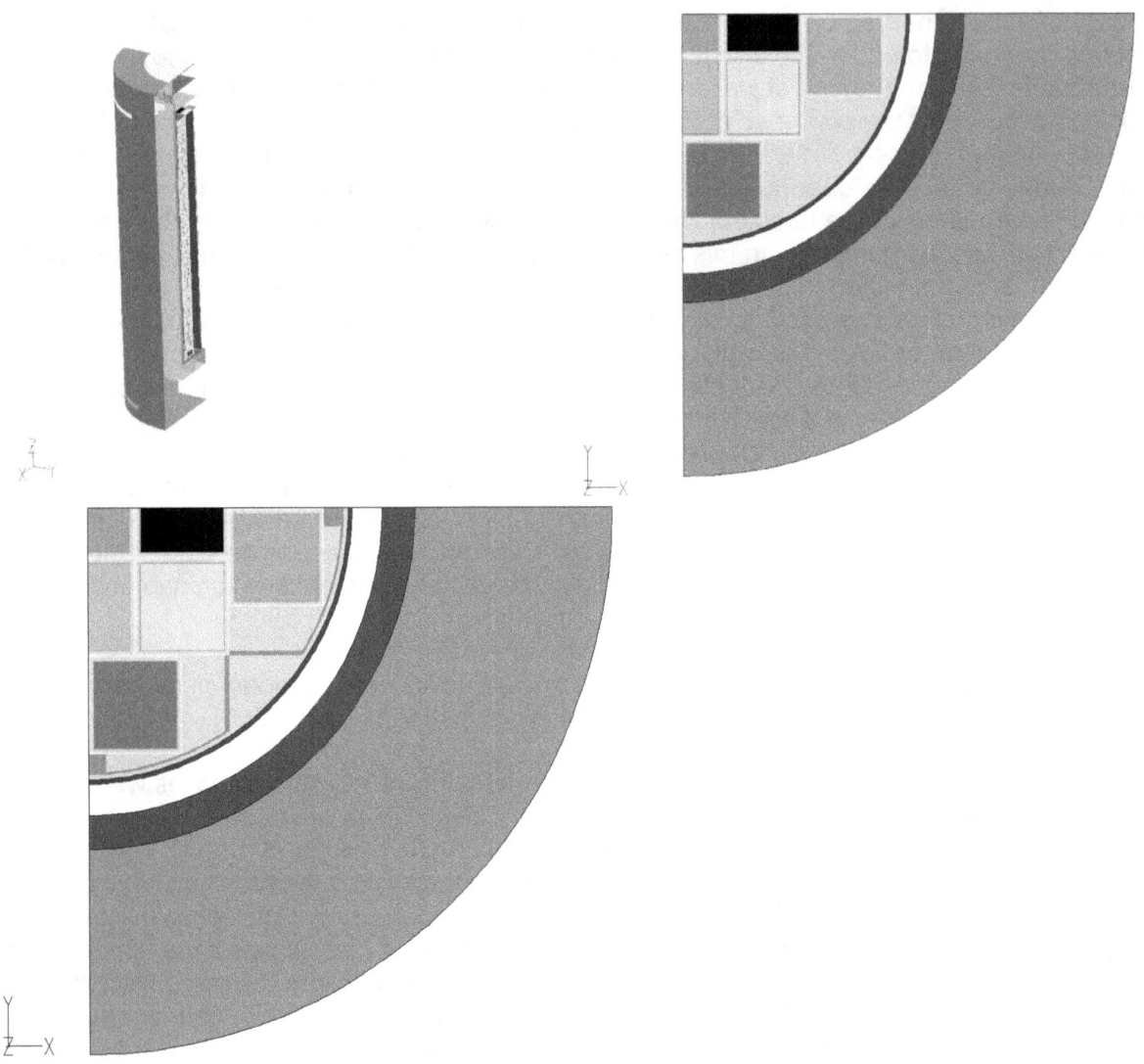

Figure 7-4 Geometry of VSC-17 dry cask

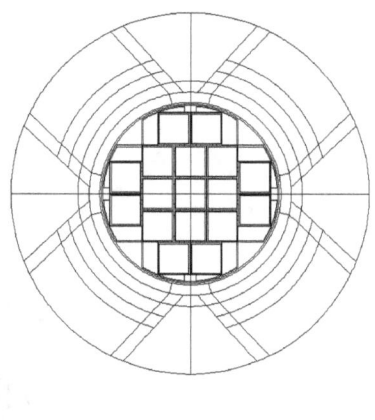

Figure 7-5 Cross section of VSC-17 dry cask

Figure 7-6 Extended control volume

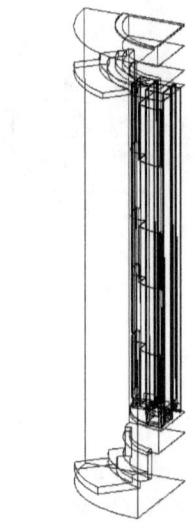

Figure 7-7a VSC-17 configuration without extended control volume

Figure 7-7b Air and helium control volume in VSC-17 without extended control volume

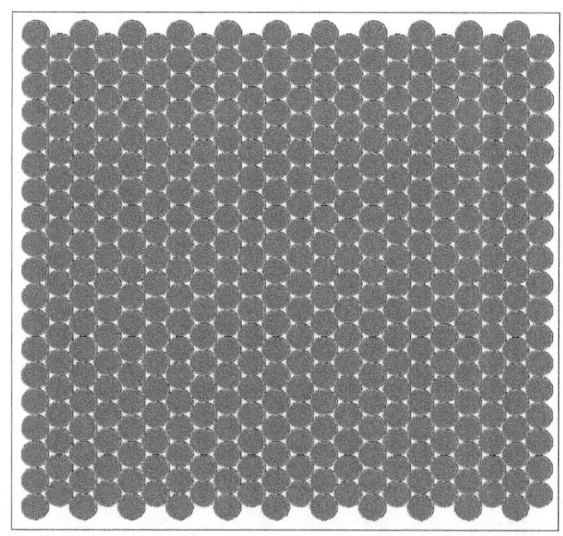

Figure 7-8a VSC-17 consolidated fuel assembly (used to calculate k$_{eff}$)

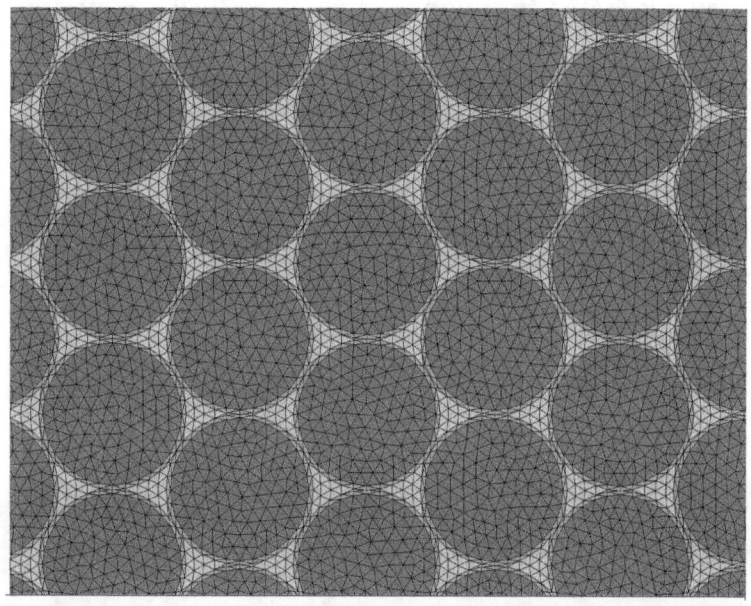

Figure 7-8b Mesh used in VSC-17 consolidated fuel assembly (used to calculate k$_{eff}$)

87

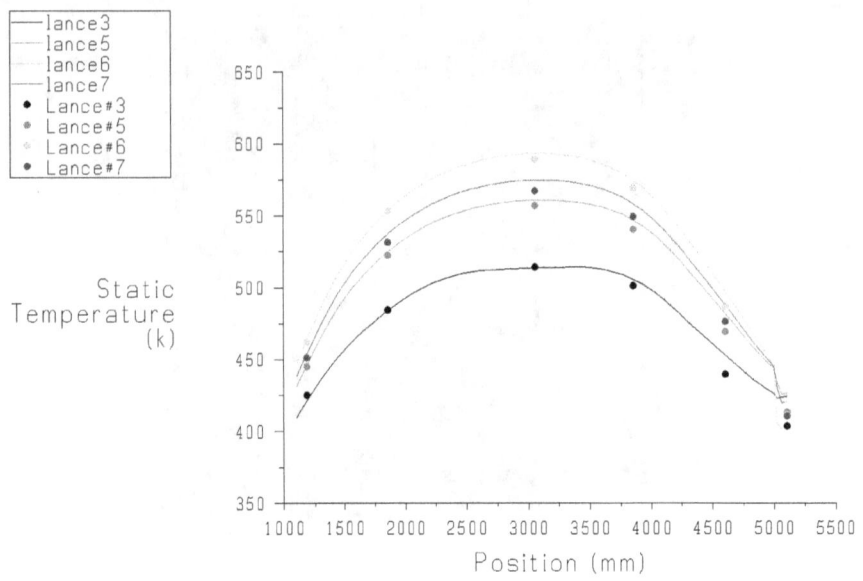

Figure 7-9a Fuel axial temperature for test #1 using extended geometry using transitional k-ω SST turbulence model (——CFD, ● Experiment)

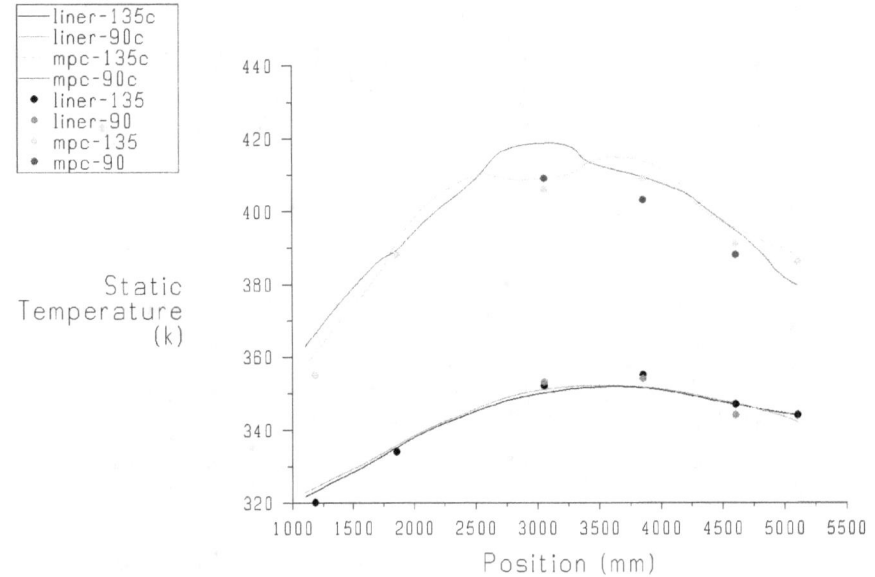

Figure 7-9b Liner and MSB walls axial temperature for test #1 using extended geometry using transitional k-ω SST turbulence model (——CFD, ● Experiment)

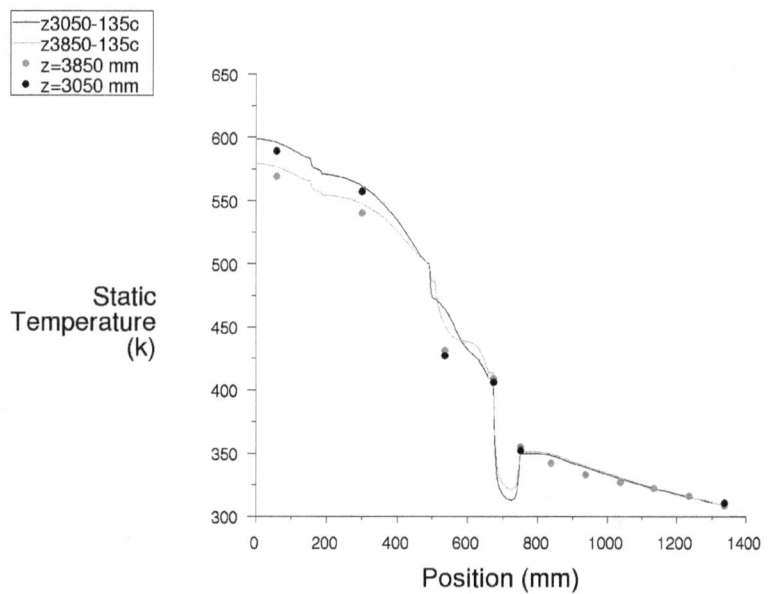

Figure 7-9c Radial temperatures at two axial locations for test #1 using extended geometry using transitional k-ω SST turbulence model (——CFD, ● Experiment)

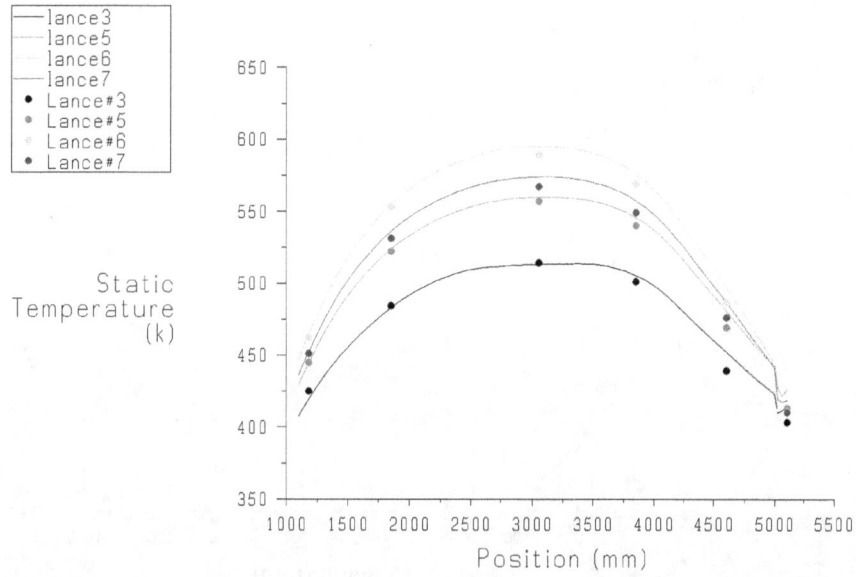

Figure 7-10a Fuel axial temperature for test #1 using extended geometry with finer grid using transitional k-ω SST turbulence model (——CFD, ● Experiment)

89

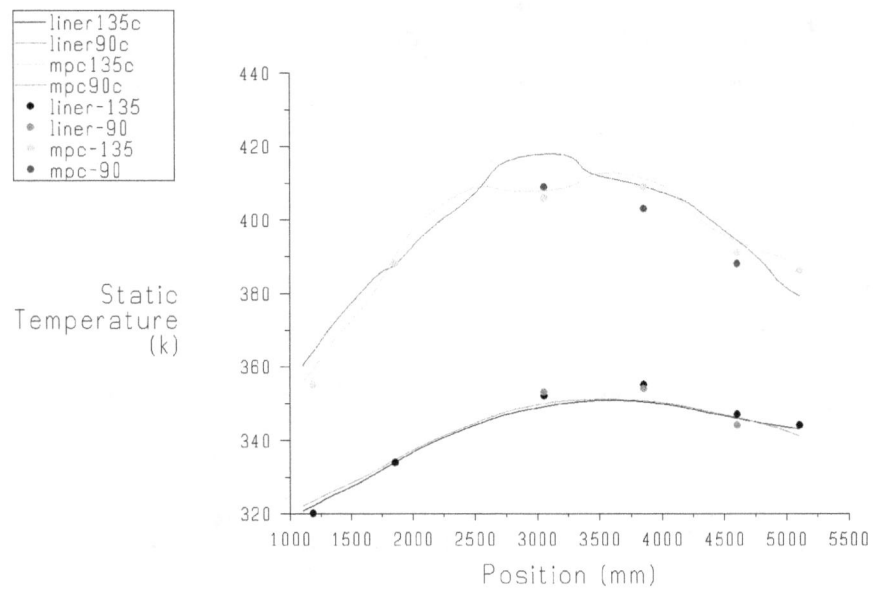

Figure 7-10b Liner and MSB walls axial temperature for test #1 using extended geometry with finer grid using transitional k-ω SST turbulence model (——CFD, ● Experiment)

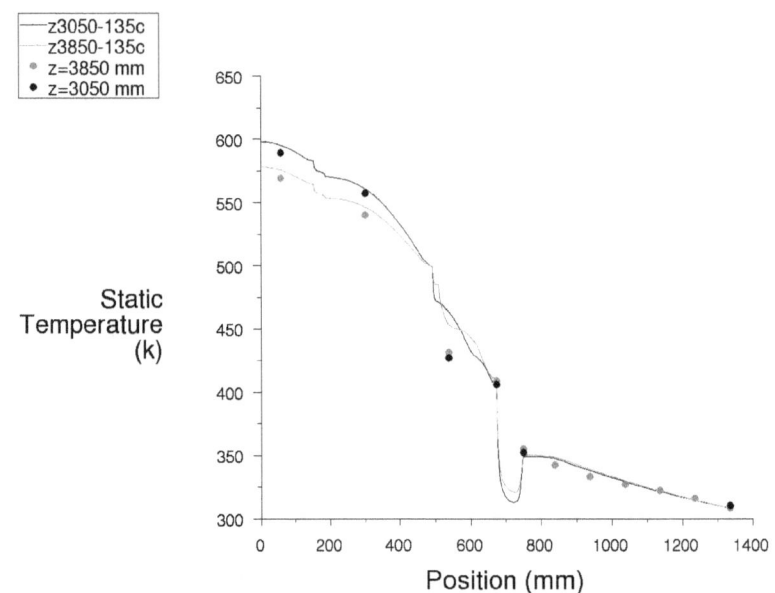

Figure 7-10c Radial temperatures at two axial locations for test #1 using extended geometry with finer grid using transitional k-ω SST turbulence model (——CFD, ● Experiment)

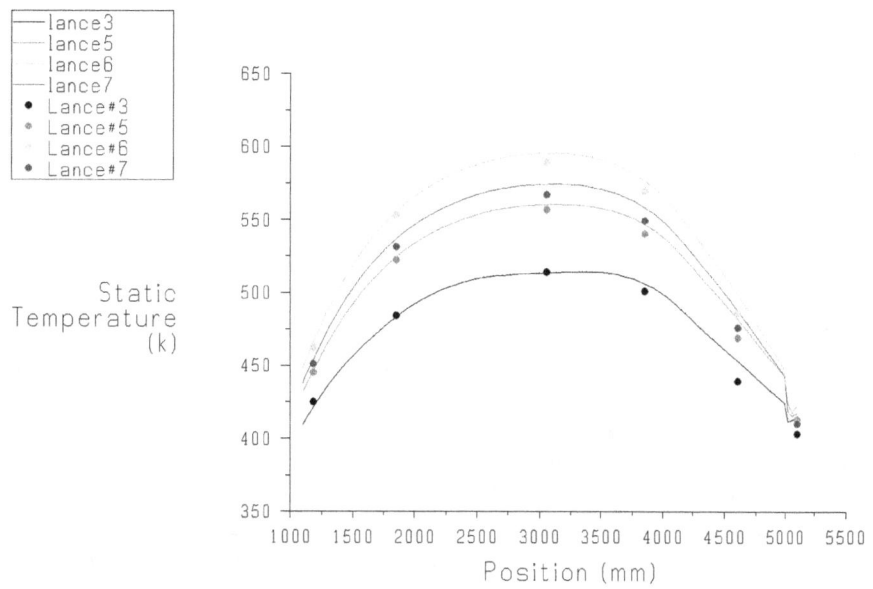

Figure 7-11a Fuel axial temperature for test #1 using transitional k-ω SST turbulence model (──CFD, • Experiment)

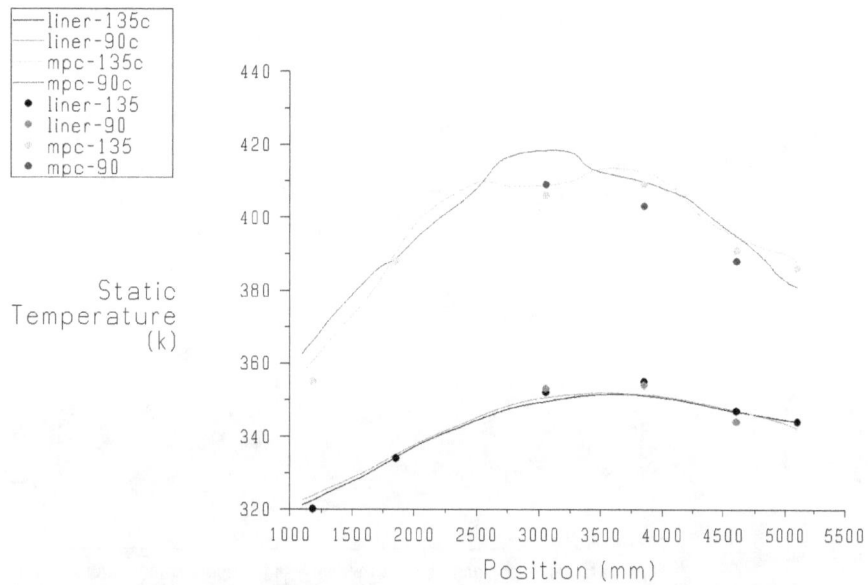

Figure 7-11b Liner and MSB walls axial temperature for test #1 using transitional k-ω SST turbulence model (──CFD, • Experiment)

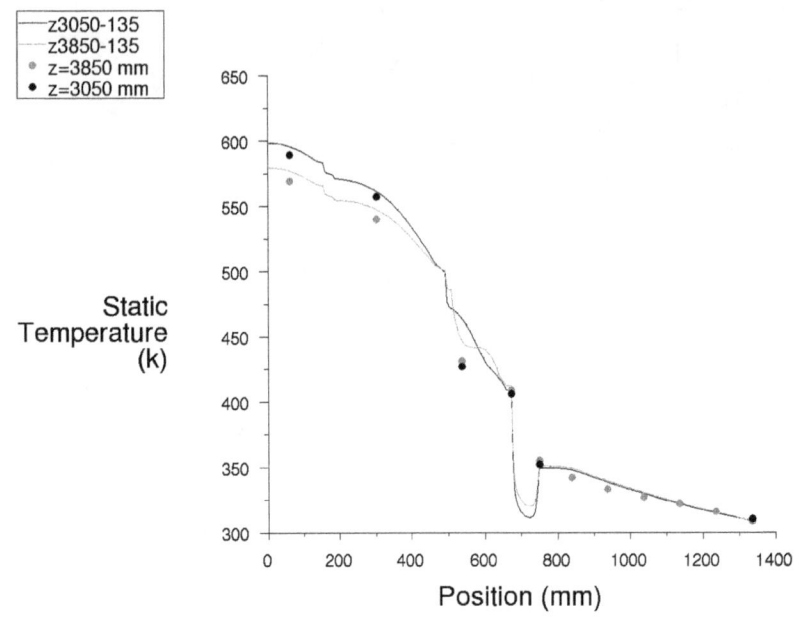

Figure 7-11c Radial temperature at two axial locations for test #1 using transitional k-ω SST turbulence model (——CFD, • Experiment)

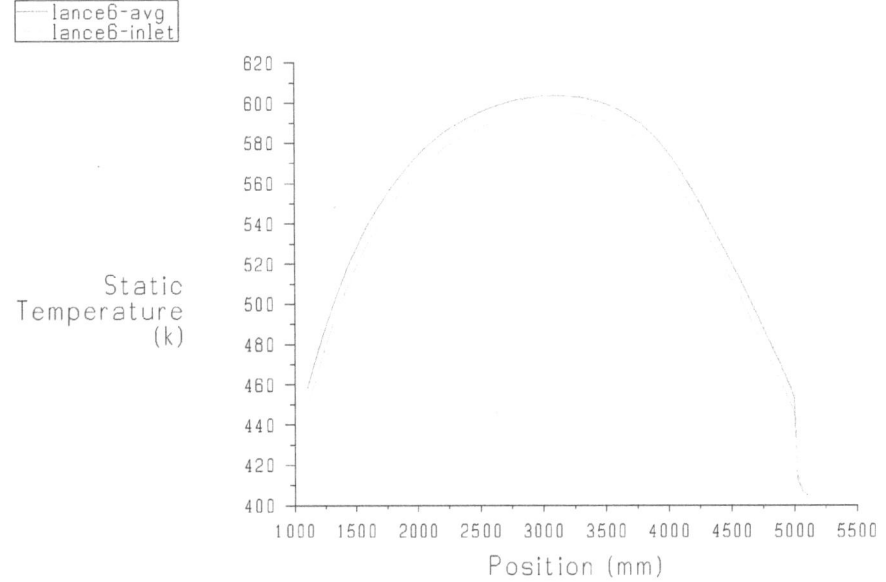

Figure 7-12a Fuel axial temperature for test #1 using transitional k-ω SST turbulence model for different operating densities (——CFD)

92

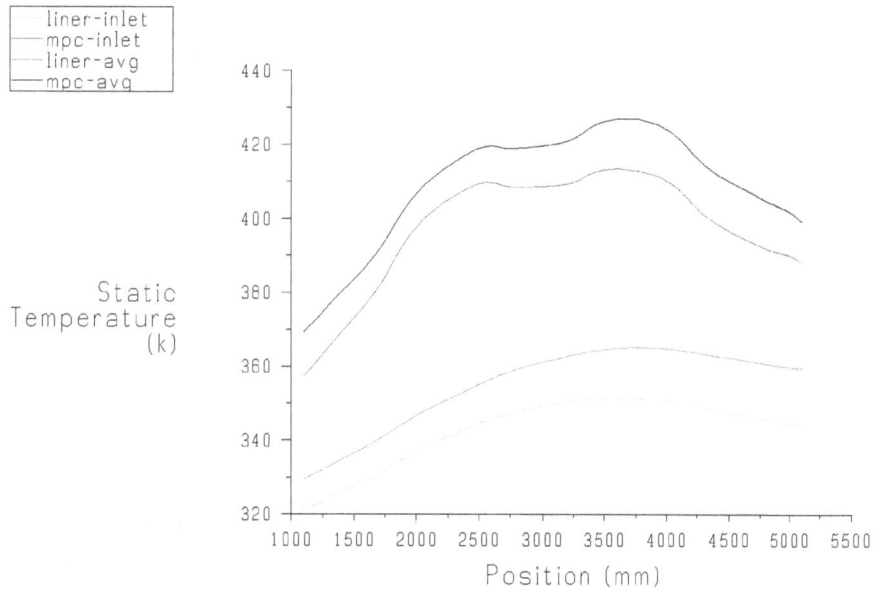

Figure 7-12b Liner and MSB walls axial temperature for test #1 using transitional k-ω SST turbulence model for different operating densities (——CFD)

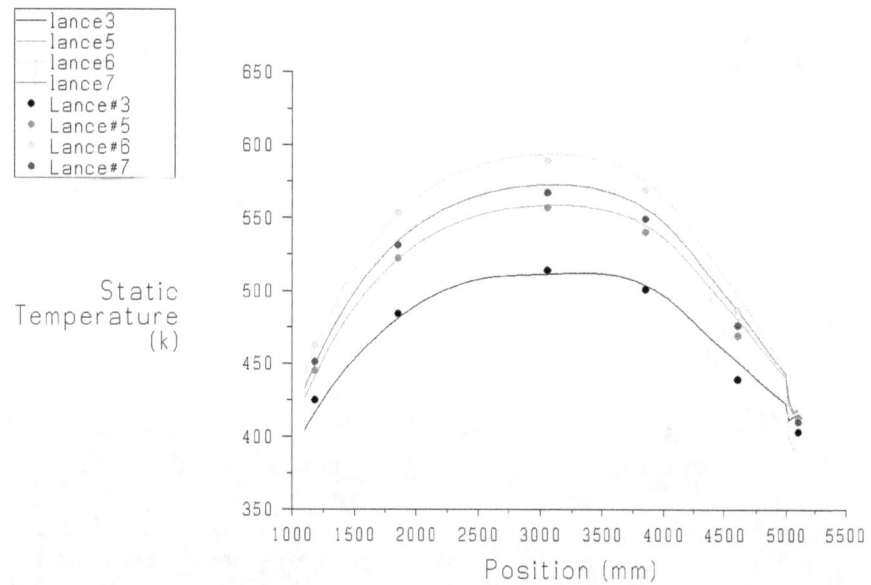

Figure 7-13a Fuel axial temperature for test #1 using low Reynolds k-ε turbulence model (——CFD, • Experiment)

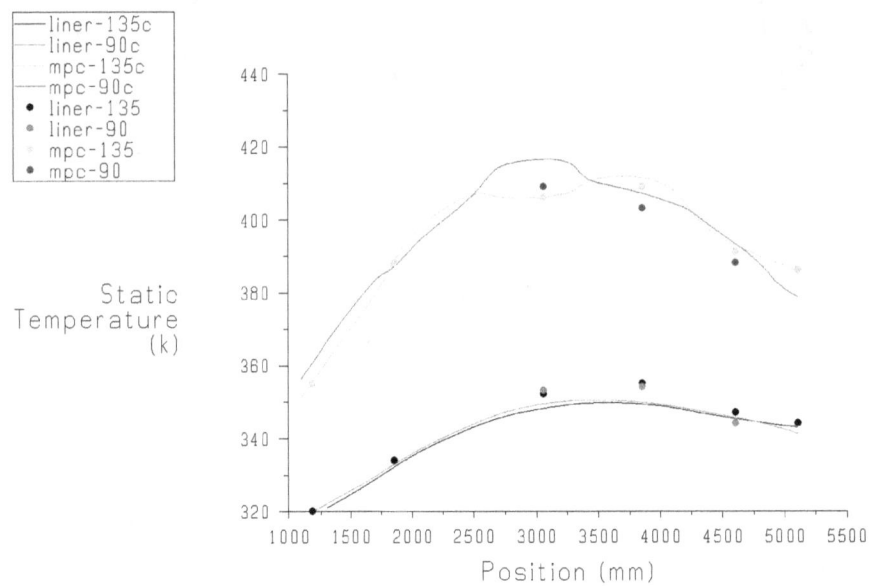

Figure 7-13b Liner and MSB walls axial temperature for test #1 using low Reynolds k-ε turbulence model (——CFD, ● Experiment)

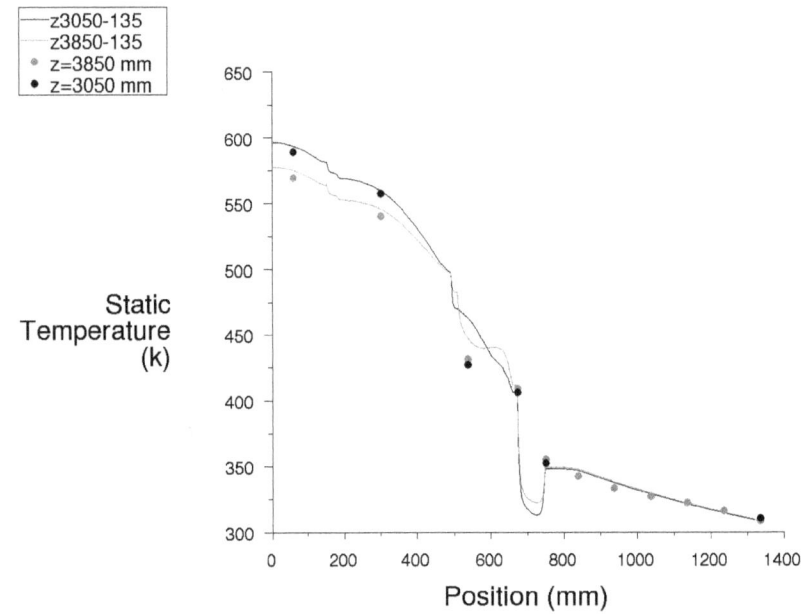

Figure 7-13c Radial temperature at two axial locations for test #1 using low Reynolds k-ε turbulence model (——CFD, ● Experiment)

94

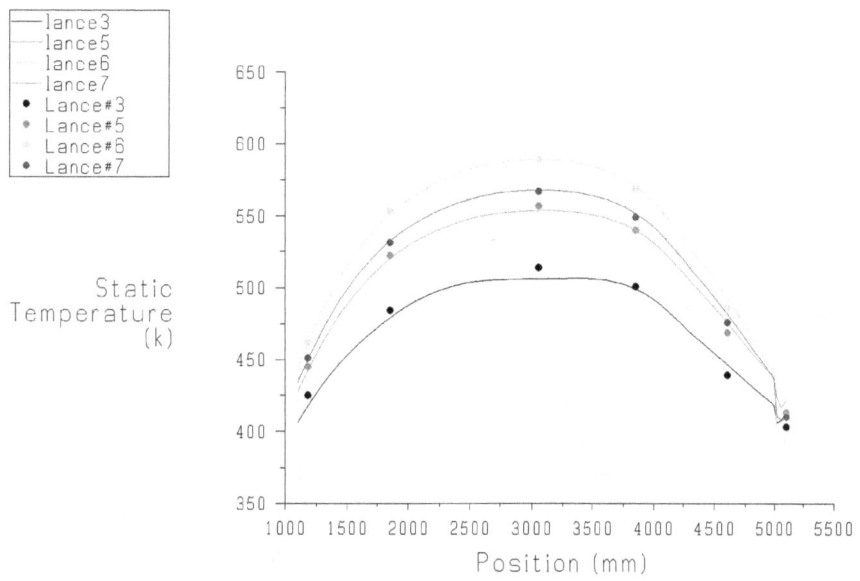

Figure 7-14a Fuel axial temperature for test #1 using standard k-ε turbulence model (——CFD, • Experiment)

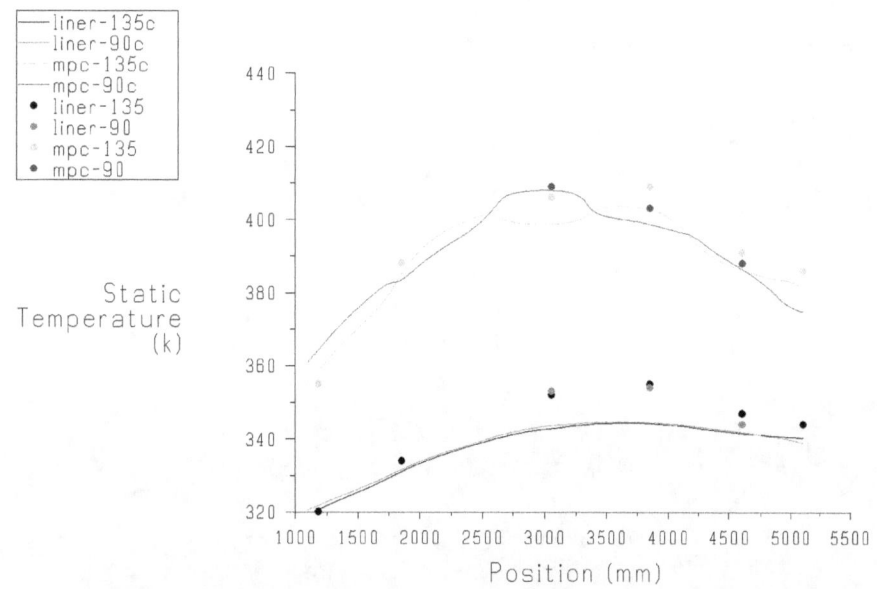

Figure 7-14b Liner and MSB walls axial temperature for test #1 using standard k-ε turbulence model (——CFD, • Experiment)

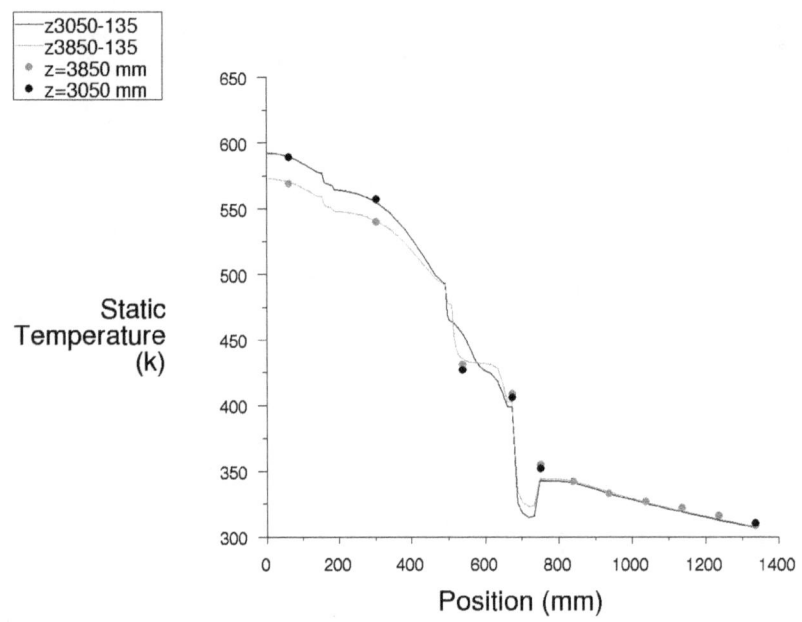

Figure 7-14c Radial temperature at two axial locations for test #1 using standard k-ε turbulence model (——CFD, ● Experiment)

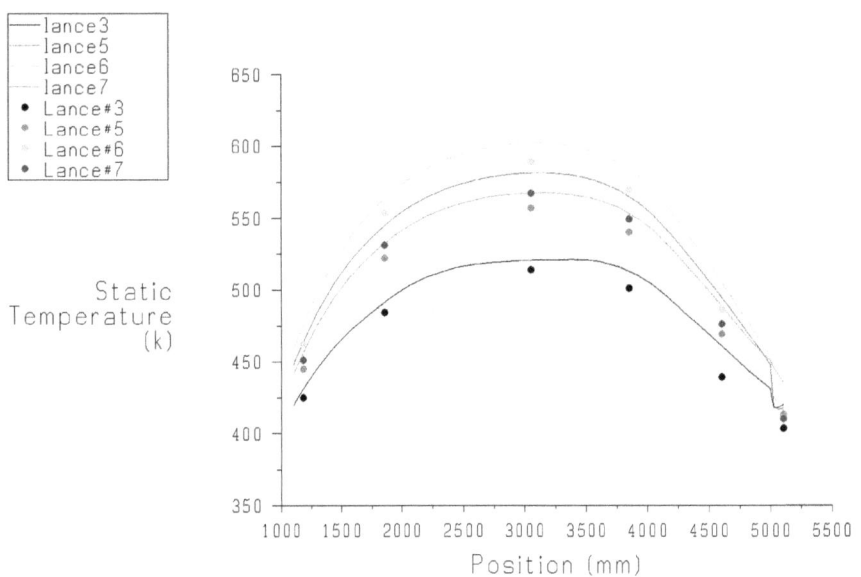

Figure 7-15a Fuel axial temperature for test #1 using laminar flow regime (——CFD, ● Experiment)

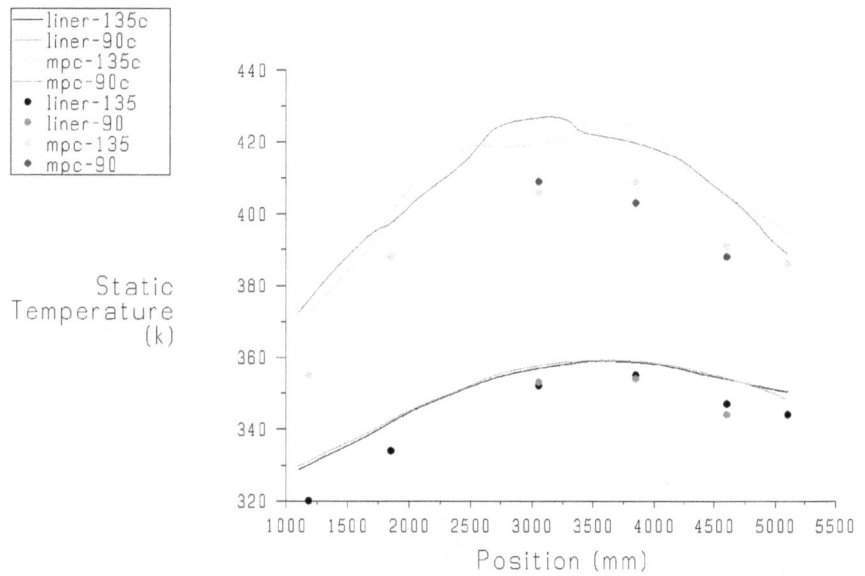

Figure 7-15b Liner and MSB walls axial temperature for test #1 using laminar flow regime (——CFD, • Experiment)

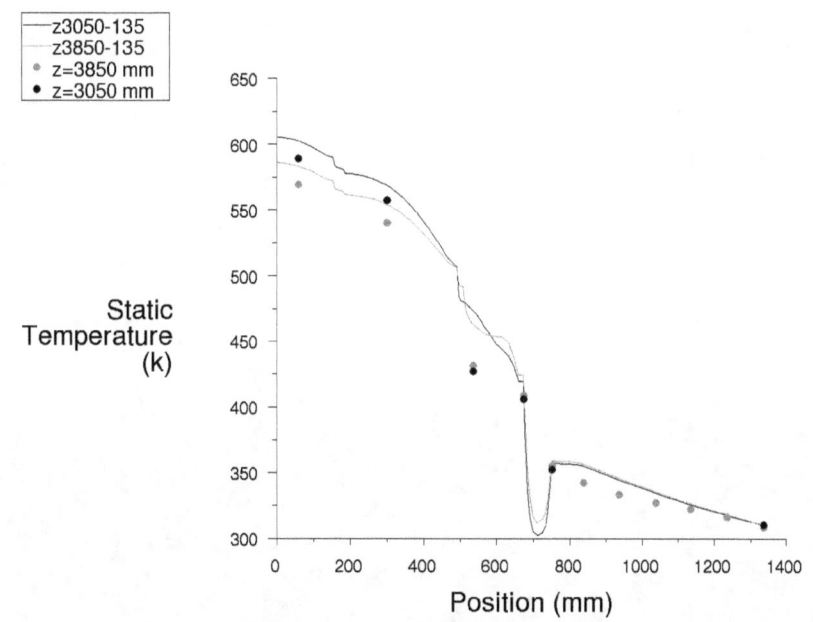

Figure 7-15c Radial temperature at two axial locations for test #1 using laminar flow regime (——CFD, • Experiment)

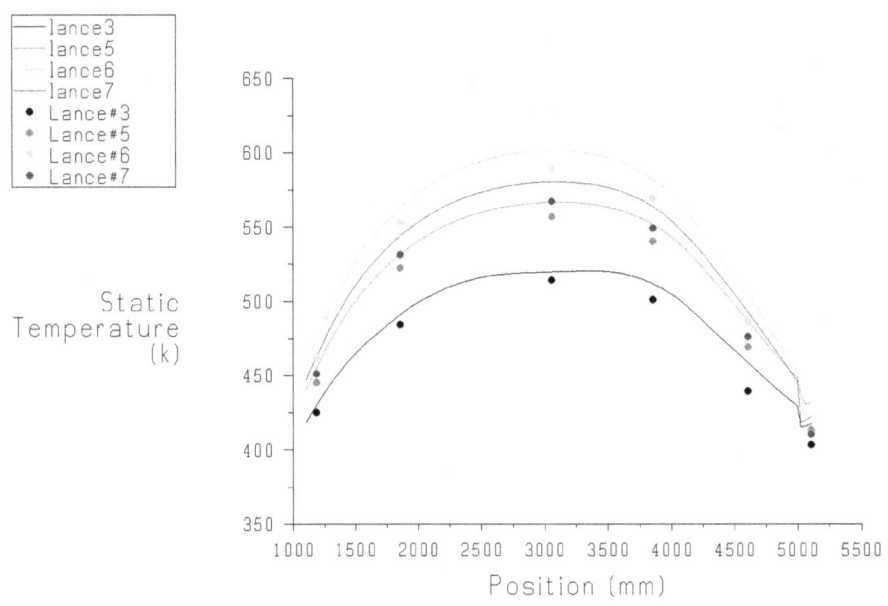

Figure 7-16a Fuel axial temperature for test #1 using transitional k-ω SST turbulence model using higher inlet temperature (——CFD, ● Experiment)

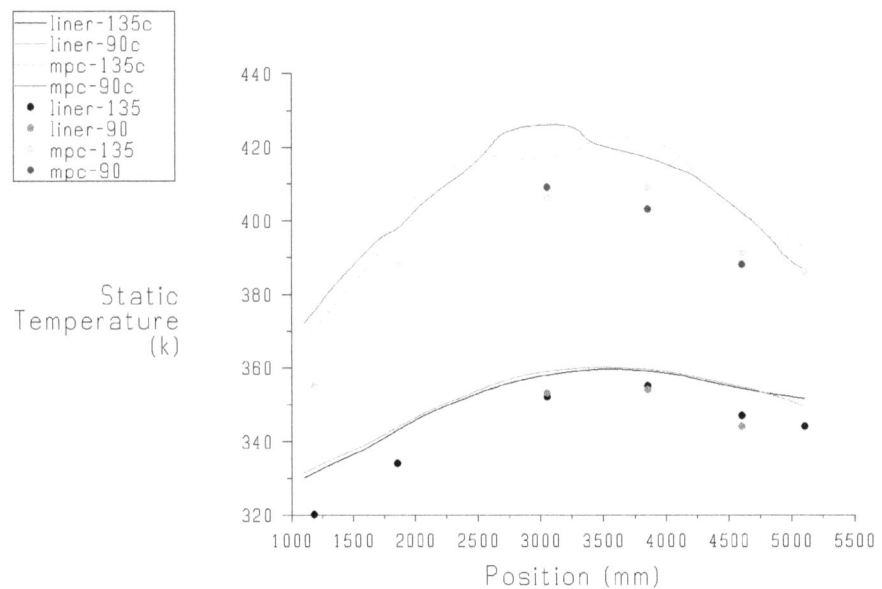

Figure 7-16b Liner and MSB walls axial temperature for test #1 using transitional k-ω SST turbulence model using higher inlet temperature (——CFD, ● Experiment)

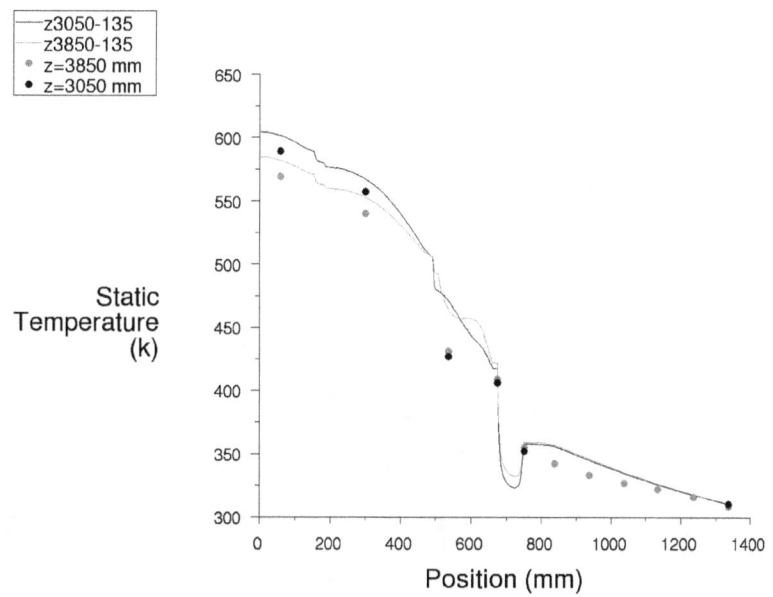

Figure 7-16c Radial temperature at two axial locations for test #1 using transitional k-ω SST turbulence model using higher inlet temperature (——CFD, • Experiment)

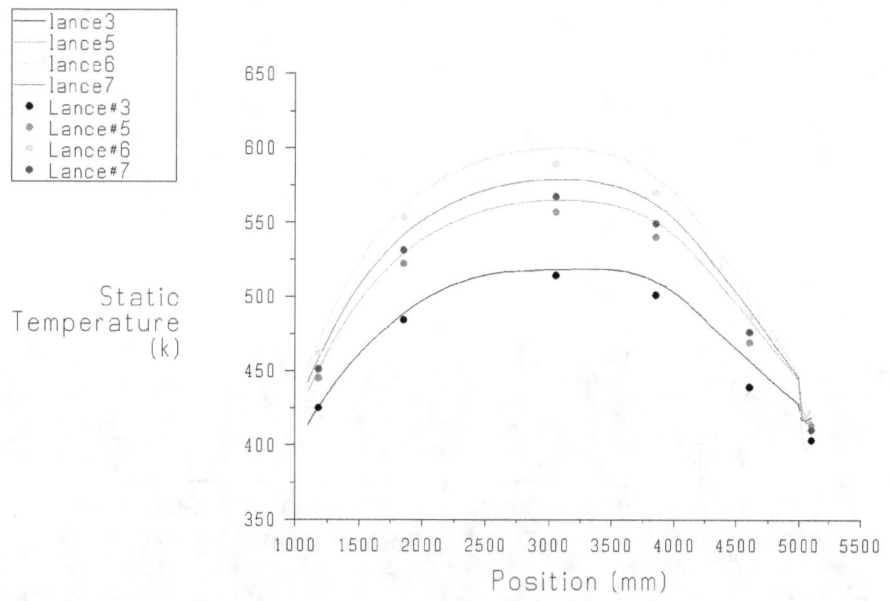

Figure 7-17a Fuel axial temperature for test #1 using low Reynolds k-ε turbulence model using higher inlet temperature (——CFD, • Experiment)

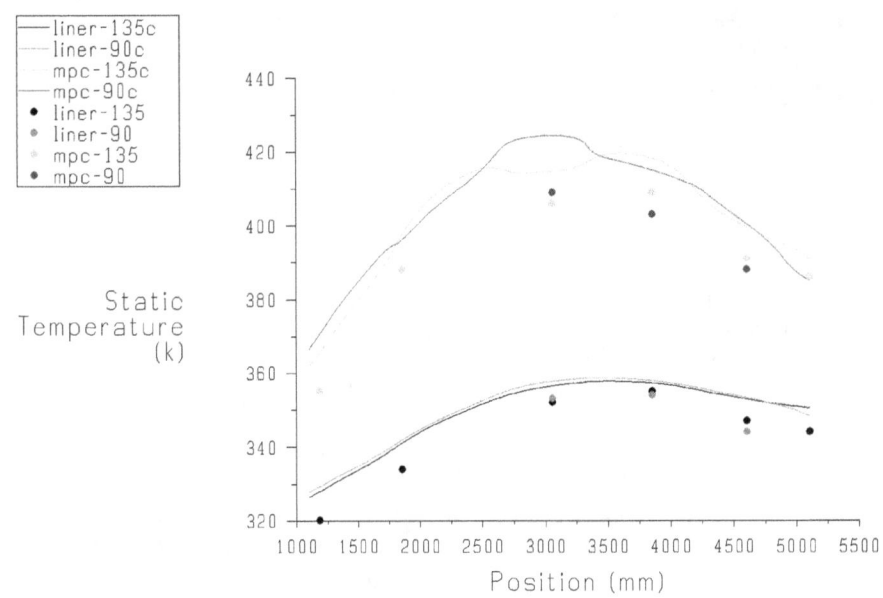

Figure 7-17b Liner and MSB walls axial temperature for test #1 using low Reynolds k-ε turbulence model using higher inlet temperature (——CFD, ● Experiment)

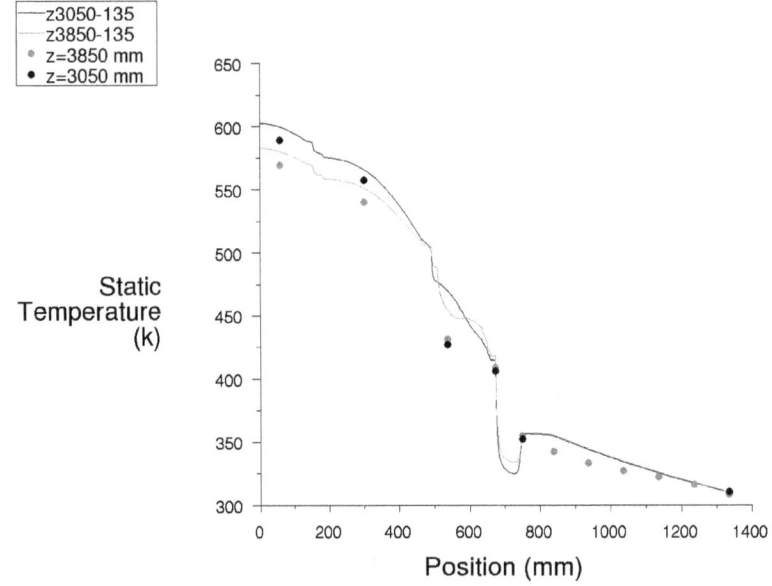

Figure 7-17c Radial temperature at two axial locations for test #1 using low Reynolds k-ε turbulence model using higher inlet temperature (——CFD, ● Experiment)

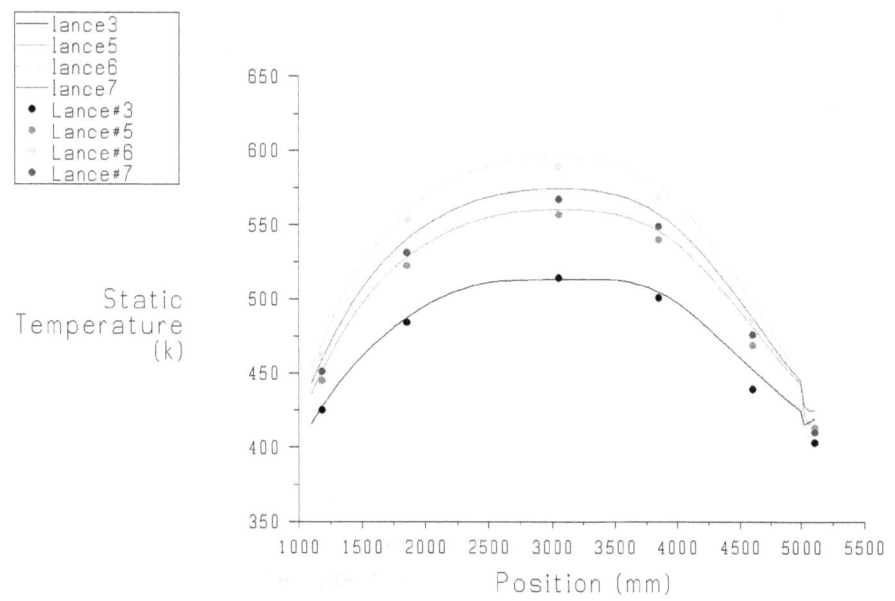

Figure 7-18a Fuel axial temperature for test #1 using standard k-ε turbulence model using higher inlet temperature (——CFD, ● Experiment)

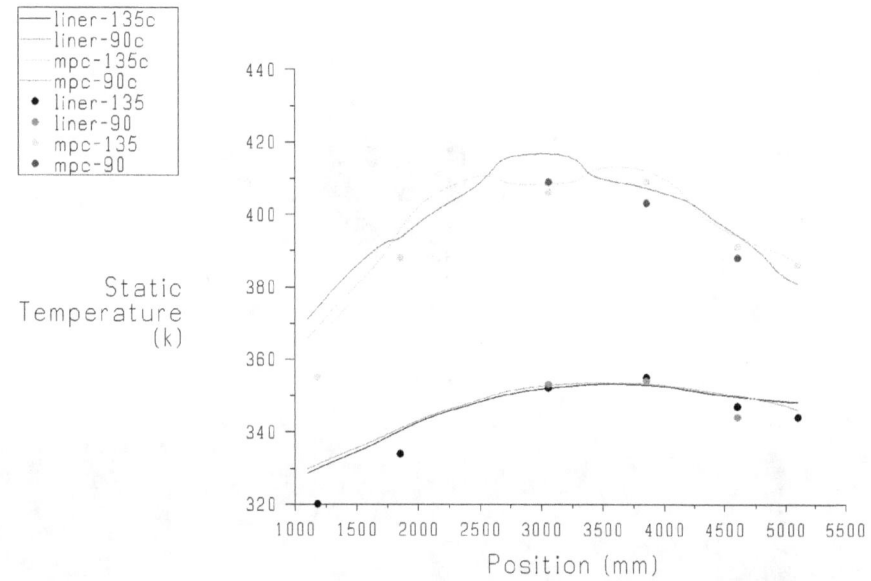

Figure 7-18b Liner and MSB axial temperature for test #1 using standard k-ε turbulence model using higher inlet temperature (——CFD, ● Experiment)

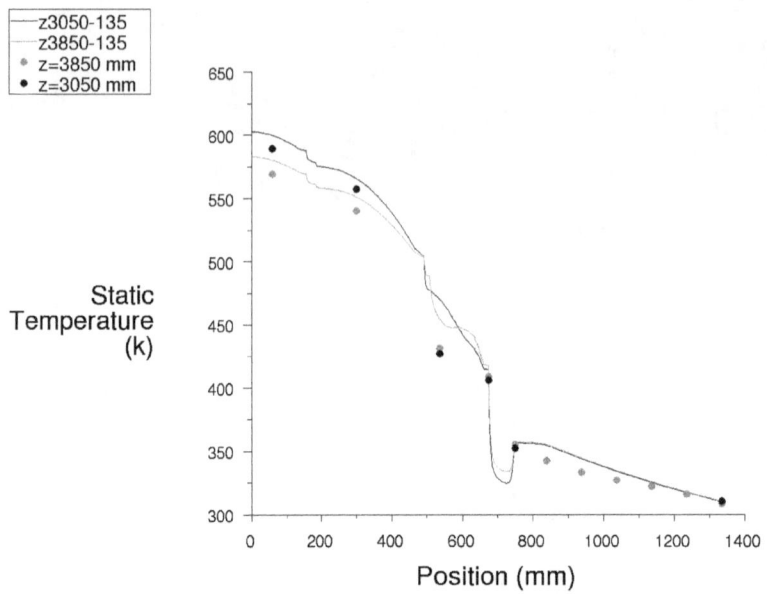

Figure 7-18c Radial temperature at two axial locations for test #1 using standard k-ε turbulence model using higher inlet temperature (——CFD, • Experiment)

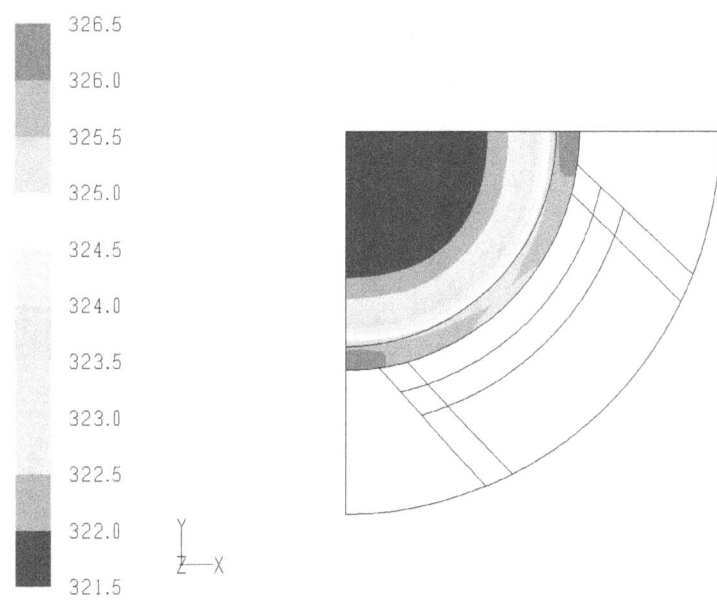

Figure 7-19 Weather cover temperature contours (K) for test #1 using low Reynolds k-ε turbulence model

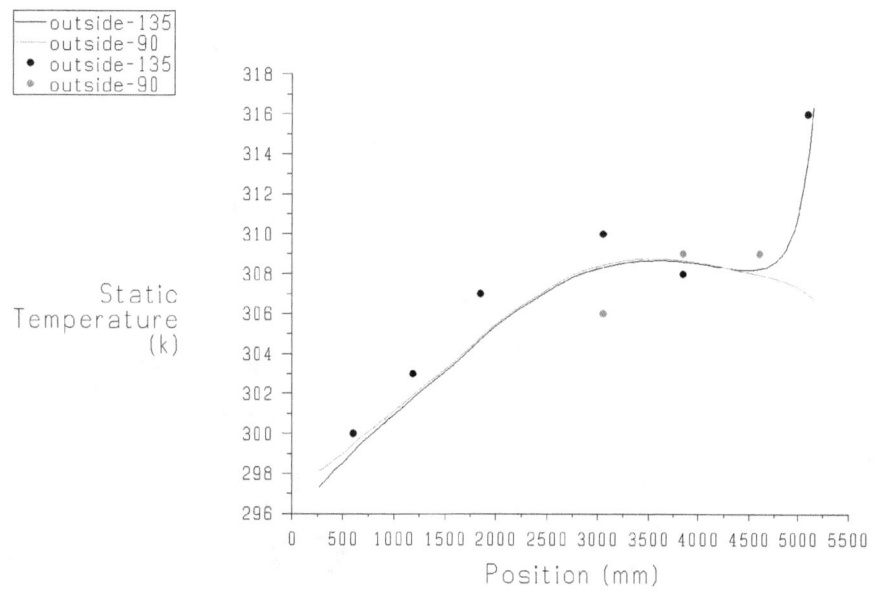

**Figure7-20 Outside surface temperature for test #1 using low Re k-ε turbulence model
(——CFD, ● Experiment)**

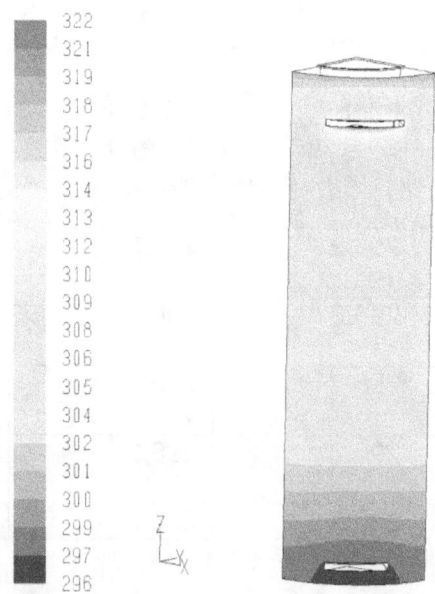

**Figure 7-21 Outside surface temperature contours (K) for test #1 using low Reynolds
k-ε turbulence model (——CFD, ● Experiment)**

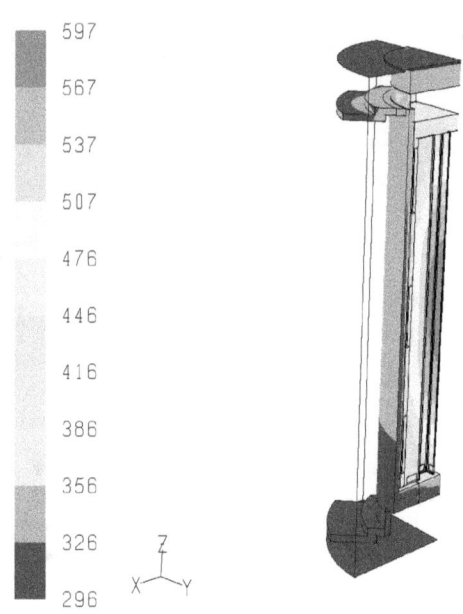

Figure 7-22 Temperature contours (K) for air and helium for test #1 using low Reynolds k-ε turbulence model

Table 7-3 Inlet and Outlet Air Temperature for Different Turbulence Models for Test #1

Control Volume	Turbulence Model	Inlet Average Temperature (K)	Inlet Max Temperature (K)	Exit Average Temperature (K)	Exit Max Temperature (K)
Dry cask + Ambient	Transitional k-ω SST	296.2	296.8	335	340
Dry cask	Transitional k-ω SST	296.2	296.2	335	339
Dry Cask	Low Reynolds k-ε	296.2	296.2	335	339
Dry cask	Standard k-ε	296.2	296.2	333	338
Dry cask	Laminar	296.2	296.2	335	352

Table 7-4 Inlet and Outlet Air Temperature for Different Turbulence Models for Test #1 Using Higher Inlet Temperature

Control Volume	Turbulence Model	Inlet Temperature (K)	Inlet Max Temperature (K)	Exit Average Temperature (K)	Exit Max Temperature (K)
Dry cask	Transitional k-ω SST	308	308	345	350
Dry cask	Low Reynolds k-ε	308	308	345	349
Dry cask	Standard k-ε	308	308	343	348

Table 7-5 VSC-17 Model Using Different Turbulence Models for Test #1

Control Volume	Turbulence Model	Air Mass Flow Rate (kg/sec)	PCT (K)	Heat Absorbed by Air (Watts)	Air Exit Temperature (K)
Dry cask + Ambient	Transitional k-ω SST	0.235	599	9,300	335
Dry cask	Transitional k-ω SST	0.238	599	9,290	335
Dry Cask	Low Reynolds k-ε	0.24	597	9,400	335
Dry cask	Standard k-ε	0.244	592	9,710	333
Dry cask	Laminar	0.201	606	8,630	335

Table 7-6 3-D VSC-17 Model Using Different Operating Density for Test #1

Test#	Operating Density (kg/m3)	PCT (K)	Air Mass Flow Rate (kg/s)	Heat Absorbed by Air (Watts)	Air Exit Temperature (K)
Test #1	1 (inlet)	598	0.238	9,284	339
Test#1	0.92 (average)	607	0.1272	7,816	351

Table 7-7 Fuel Radial k_{eff} for VSC-17 with Helium

Temperature (K)	K_{eff} (W/(m-K))
370	1.383
509	1.770
647	2.01
676	1.99
703	1.85

Table 7-8 Fuel Axial k_{eff} for VSC-17 with Helium

Temperature (K)	Fuel with Helium (W/(m-K))
366	5.75
505	5.84
644	4.99
673	4.75
720	4.36

7.2   Conclusions

Temperature measurements for the VSC-17 spent fuel storage cask undertaken by INL were used to validate the FLUENT 3-D CFD model. From this validation, the following was concluded:

- The flow in the air channel is found to be in the transitional region of turbulence. Only turbulence models that are able to deal with this region of the flow regime should be used to analyze this type of flow.

- Among the available turbulence models in FLUENT, the transitional k-ω SST and the low Reynolds k-ε models were able to predict the experimental data.

- The low Reynolds k-ε model is preferred because it includes the effect of gravity in the production and dissipation of turbulent kinetic energy.

- These turbulence models require finer mesh near the wall. The dimensionless distance from the wall to the center cell of the first row of cells (y^+) should be of the order of unity.

- The laminar flow regime overpredicts the PCT and is not appropriate to analyze the air flow.

- The standard k-ε model is not suitable to model the air flow. This model overpredicts air flow heat transfer.

- The flow inside the MSB is laminar. Only laminar flow is appropriate to model flow inside the MSB.

- The K_{eff} model for the spent fuel region successfully matched the experimental data, thus making it an appropriate assumption to model the fuel rods region.

- When the inlet and outlet ducts to the dry cask are used as pressure boundaries, the specified operating density must be that corresponding to the ambient inlet conditions.

- For higher elevations both ambient pressure decreases and air mass flow rate decrease. Subsequently, PCT will increase.

8.0   BRIEF BPGs CHECKLIST

Initial Preparation

- Is the problem described clearly?
- Is the modeled control volume or system specified?
- Is the scenario to be modeled well defined?
- Are the study objectives clearly listed?

Geometry Generation

- Is the coordinate system correct?
- Are the units correct?
- Have any substantial modifications been made to the geometry?
- Is the geometry complete?
- Are there oversimplifications because of symmetry assumptions, etc.?

Grid Generation

- Are the grid angles larger than 40 degrees and less than 140 degrees?

- Are the ratios of adjacent volumes less than 2?

- Are the aspect ratios below the values given in the solver manual (typically, 10–50)?

- Is the grid scalable (keeping y-plus and t-plus similar between scaled experiment and prototypic design)?

- Are grid nodes concentrated in areas of foreseeable physical significance?

- Does the grid contain nonmatching grid interfaces in critical regions?

- Is the grid compatible with the physical models (turbulence model, wall treatments, etc.)?

- Are inlet, outlet, symmetry, and cyclic boundary condition regions located correctly?

Selection of Physical Models

- Are the prevalent physical phenomena and the features of the flow field well understood?

- Is one dealing with laminar, transitional, or turbulent flow?

- If unsure of the flow mode, a conservative approach (i.e., laminar mode) should be used.

- If laminar, make sure that a finer grid at the wall is used to resolve the boundary layers.

- If transitional, the following should be taken into account:

  o   Was the correct turbulence model selected?

109

- If required, was a finer grid used at the wall?

- Does the choice of turbulence level adequately represent the study objective? (Choices are: RANS, T-RANS, LES, or hybrid approach.)

- Is the selected statistical model to represent turbulence appropriate? (This is valid if RANS or T-RANS were selected.)

- Is the selected wall turbulence model appropriate with the flow features?

- Is the selected wall turbulence model appropriate with the selected turbulence model (standard wall function only should be used with standard k-ε)?

- If porous media model was used, the following should be taken into account:

  - Were the correct inertial and frictional loss coefficients used? (See Appendix B for additional information on how to evaluate the inertial and frictional loss coefficient factors).

  - Were the correct units used for the inertial and frictional coefficient as prescribed in the CFD software?

  - Was the correct porosity used to calculate the effective thermal conductivity?

  - Was the correct effective thermal conductivity used to represent radiation and conduction heat transfer?

  - Is the porous media model more or less conservative than the real physical model?

  - Was the emissivity set to zero at the walls surrounding the porous media volume?

- Are the boundary conditions consistent with the choice of turbulence model?

- Are the boundary conditions consistent with the surrounding and the physics of the problem?

- Was a sensitivity analysis performed for boundary placement?

- If a pressure boundary condition is used, the following should be taken into account:

  - Was the correct operating density selected?

  - Does the operating density correspond to the inlet pressure and temperature?

- What equation of state was used to describe the density of the fluids? (For helium and air, ideal gas state equation should be used.)

- For ventilated casks, were losses at the screen of the inlet and outlet vents accounted for?

- Were thermophysical properties for the fluids as a function of temperature implemented correctly?

- If kinetic theory was used to calculate thermophysical values for fluid, was the Leonard Jones constant for each gas checked?

- Were thermophysical properties for the solids and walls as a function of temperature implemented correctly?

- Were the correct emissivity values used at the walls?

- Was a target used to check for convergence?

- Was overall mass and energy balance checked?

- Was the radiation heat transfer model used adequate for the type of problem?

- Was a sensitivity analysis on the radiation heat transfer model input performed?

- If an axisymmetric model is used, was the model more conservative than a 3-D representation of the cask?

Numerical Methods

- Was the first-order upwind spatial discretization avoided unless necessary (to avoid numerical diffusion)?

- Was the first-order implicit time integration avoided unless necessary (to avoid numerical diffusion)?

- If LES is used, was a higher-order central differencing scheme selected (preferably fourth order)?

Verification

- Were round-off errors checked (by comparing single to double precision results)?

- Were errors associated with selection of iteration convergence criteria checked?

- Were errors associated with space discretization checked (as shown in Appendix A)?

- Were errors associated with time discretization checked (as shown in Appendix A)?

- To verify the solution, use the GCI method (Appendix A) for at least four grids to compute and compare the degree of accuracy as explained in ASME (2009).

- GCI method should meet the following criteria (ASME, 2009):

  o Systematic grid refinement in all the directions should be performed.
  o Observed and expected order of accuracy should be comparable.

111

- o   Grid resolutions should be in the asymptotic region.
- o   If any of these conditions are not met, conservative order of accuracy should be used to calculate the discretization error.

- Were procedures to limit and locate user errors followed (including selection of high-quality user interface to the CFD code and the use of quality assurance practices)?

Validation

- Follow a tiered approach comparing first to separate effects experiments (unit problems) and working up through complete system experiments.

- Use repeat experiments when possible to help quantify experimental error.

- Select target variables and metrics for agreement between calculation and experiment.

- Characterize experimental uncertainty for all target variables, distinguishing between random and systematic (bias) contributions to the uncertainty.

- Perform uncertainty analysis on the simulation, if sufficient computer resources are available, to place bounds on results.

REFERENCES

American Institute of Aeronautics and Astronautics (AIAA), "AIAA Guide for the Verification and Validation of Computational Fluid Dynamics Simulations," AIAA G-077-1998.

American Society of Mechanical Engineers (ASME), "Standard for Verification and Validation in Computational Fluid Dynamics and Heat Transfer," V&V 20-2009.

Baldwin, W.S. and H. Lomax, "Thin-Layer Approximation and Algebraic Model for Separated Turbulent Flows," AIAA Paper 78-257, 1978.

Bartzis, J.G. and D. Vlachogiannis, A. Sfetsos, "Thematic Area 5: Best Practice Advice for Environmental Flows," *The QNET-CFD Network Newsletter*, Vol. 2, 34–39, 2004.

Cadafalch, J. and C.D. Pérez-Segarra, R. Cònsul, A. Oliva, "Verification of Finite Volume Computations on Steady-State Fluid Flow and Heat Transfer," *Journal of Fluids Engineering*, Vol. 124, 11–21, 2002.

Casey, M. and T. Wintergerste (eds.), "Special Interest Group on 'Quality and Trust in Industrial CFD' Best Practice Guidelines, Version 1," ERCOFTAC Report, 2000.

Celik, I. and A. Badeau, Jr., "Verification and Validation of DREAM Code," Report No. MAE_IC03/TR103, Mechanical and Aerospace Engineering Department, West Virginia University, Morgantown, WV, 2003.

Celik, I., "Procedure for Estimation and Reporting of Discretization Error in CFD Applications," Mechanical and Aerospace Engineering Department, West Virginia University, Morgantown, WV, 2005.

Celik, I. and O. Karatekin, "Numerical Experiments on Application of Richardson Extrapolation With Nonuniform Grids," *Journal of Fluids Engineering*, Vol. 119, 584-590, 1997.

Coleman, H. and F. Stern, A. Di Mascio, E. Campana, "The Problem with Oscillatory Behaviour in Grid Convergence Studies," *Journal of Fluids Engineering*, Vol. 123, 438–439, 2001.

Coleman, H. W. and F. Stern, "Uncertainties and CFD Code Validation," *Journal of Fluids Engineering*, Vol. 119, 795–803, 1997.

Ferziger, J.H. and M. Perić, *Computational Methods for Fluid Dynamics*, Springer-Verlag, Berlin Heidelberg, 1999.

Ferziger, J.H. and M. Perić, *Computational Methods for Fluid Dynamics*, Springer-Verlag, Berlin Heidelberg New York, 3rd edition, 2002.

Fisher, E.H. and N. Rhodes, "'Uncertainty in Computational Fluid Dynamics,' Proceedings of the Institution of Mechanical Engineers, Part C: Journal of Mechanical Engineering Science," *Journal of Mech. Eng. Sci.*, Vol. 210, 91-94, 1996.

Fletcher, C.A.J., "Computational Techniques for Fluid Dynamics," Vol. I and Vol. II," Springer-Verlag, Berlin Heidelberg New York, 2nd edition, 1991.

FLUENT User Guide Version 6, Fluent Inc., NH, September 2006.

Fothergill, C.E. and P.T. Roberts, A.R. Packwood, "Flow and Dispersion around Storage Tanks. A Comparison between Numerical and Wind Tunnel Simulations," *Wind & Structures*, Vol. 5, 89–100, 2002.

Franke, J. and A. Hellsten, H. Schlünzen, B. Carissimo, "Best Practice Guideline for the CFD Simulation of Flows in the Urban Environment," European Cooperation in Science and Technology (COST) Action 732, Belgium, 2007.

Ghosal, S. and P. Moin, "The Basic Equations for the Large Eddy Simulation of Turbulent Flows in Complex Geometry," *Journal of Computational Physics*, Vol. 118, 24–37, 1995.

Giles, M.B., "Nonreflecting Boundary Conditions for Euler Equation Calculations," *AIAA Journal*, Vol. 28, 2050–2058, 1990.

Hargreaves, D.M. and N.G. Wright, "On the Use of the k-ε Model in Commercial CFD Software to Model the Neutral Atmospheric Boundary Layer," *Journal of Wind Engineering and Industrial Aerodynamics*, Vol. 95, 355–369, 2007.

Hirt, C.W., "Heuristic Stability Theory for Finite-Difference Equations," *Journal of Computational Physics,* Vol. 2, 339–355, 1968.

Hirsch, C., *Numerical Computation of Internal and External Flows,* Vol. I and Vol. II, John Wiley & Sons Ltd., NY, 1991.

Hirsch C. and V. Bouffioux, F. Wilquem, "CFD Simulation of the Impact of New Buildings on Wind Comfort in an Urban Area," Workshop Proceedings, COST Action C14, *Impact of Wind and Storm on City Life and Built Environment,* Nantes, France, 2002.

Idelchik, I.E., "Handbook of Hydraulic Resistance," 3rd edition, CRC Press, 1993.

"JANAF Thermochemical Tables," 3rd edition, published by American Chemical Society, American Institute of Physics for the National Bureau of Standards, Vol. 14, Washington, DC, 1985.

Ince, N.S. and B.E. Launder, "Three-Dimensional and Heat-Loss Effects on Turbulent Flow in a Nominally Two-Dimensional Cavity," *International Journal of Heat and Fluid Flow*, Vol. 16, 171-177, 1995."

Knupp, P. and K. Salari, *Verification of Computer Codes in Computational Science and Engineering*, Chapman and Hall/CRC, Boca Raton, FL, 2003.

Launder, B.E. and D.B. Spalding, "The Numerical Computation of Turbulent Flow," Computer Methods in Applied Mechanics and Engineering., Vol. 3, 269–289, 1974.

McAdams W.H., *Heat Transmission*, McGraw-Hill Book Company, Inc., NY, 1954.

McKinnon M.A. and R.E. Dodge, R.C. Schmitt, L.E. Eslinger, G. Dineen, "Performance Testing and Analyses of the VSC-17 Ventilated Concrete Cask," TR-100305, Electric Power Research Institute, Palo Alto, CA, 1992.

Menter, F.R., "Zonal Two Equation k-ω Turbulence Models for Aerodynamic Flows," AIAA Paper 93-2906, 1993.

Menter, F.R., "A Comparison of Some Recent Eddy-Viscosity Turbulence Models," *Journal of Fluids Engineering*, Vol. 118, 514–519, 1996.

Menter, F., "CFD Best Practice Guidelines for CFD Code Validation for Reactor-Safety Applications," European Commission, 5th EURATOM Framework Programme, Report EVOL-ECORA-D1, 2002.

NUREG-1536, Revision 1, "Standard Review Plan for Spent Fuel Dry Cask Storage Systems at a General License Facility," U.S. Nuclear Regulatory Commission, Washington DC, 2010.

NUREG/CR-6978, "A Phenomena Identification and Ranking Table (PIRT) Exercise for Nuclear Power Plant Fire Modeling Applications", U.S Nuclear Regulatory Commission, Washington DC, 2008.

NEA/CSNI/R(2007)5, "Best Practice Guidelines for the use of CFD in Nuclear Reactor Safety Applications," Organisation for Economic Co-operation and Development, Nuclear Energy Agency, Committee on the Safety of Nuclear Installations, 2007.

Oberkampf, W.L. and T.G. Trucano, "Verification and Validation in Computational Fluid Dynamics," *Progress in Aerospace Sciences*, Vol. 38, 209–272, 2002.

Oberkampf, W.L. and T.G. Trucano, C. Hirsch, "Verification, Validation and Predictive Capability in Computational Engineering and Physics," *Applied Mechanics Review*s, Vol. 57, 345–384, 2004.

Oberkampf, W.L. and M.F. Barone, "Measures of Agreement between Computation and Experiment: Validation Metrics," *Journal of Computational Physics*, Vol. 217, 5–36, 2006.

Patel, V.C. and W. Rodi, G. Scheuerer, "Turbulence Models for Near-Wall and Low Reynolds Number Flows: A Review," *AIAA Journal*, Vol. 23, 1,308–1,319, 1985.

Richardson, L.F., "The Approximate Arithmetical Solution by Finite Differences of Physical Problems Involving Differential Equations, with an Application to the Stresses in a Masonry Dam," *Royal Society of London, Series A, Containing Papers of a Mathematical or Physical Character*, 1910.

Richardson, L.F., "The Deferred Approach to the Limit. Part I," *Philosophical Transactions of the Royal Society of London. Series A, Containing Papers of a Mathematical or Physical Character* Vol. 226, 299-361, 1927.

Rizzi, A. and J. Voss, "Toward Establishing Credibility in Computational Fluid Dynamics Simulations," *AIAA Journal*, Vol. 36, 668–675, 1998.

Roache, P.J., "Perspective: A Method for Uniform Reporting of Grid Refinement Studies, *Journal of Fluids Engineering*, Vol. 116, 405–413, 1994.

Roache, P.J., "Quantification of Uncertainty in Computational Fluid Dynamics," *Annual Review of Fluid Mechanics*, Vol. 29, 123–160, 1997.

Roache, P.J., *Verification and Validation in Computational Science and Engineering,* Hermosa Publishers, Albuquerque, NM, 1998.

Roache, P.J., "Code Verification by the Method of Manufactured Solutions, *ASME Journal of Fluids Engineering*, Vol. 124, 4–10, 2002.

Rodi, W., "Progress in Turbulence Modelling for Incompressible Flows," AIAA Paper 81–45, St. Louis, MO, 1981.

Roy, C.J., "Review of Code and Solution Verification Procedures for Computational Simulation, *Journal of Computational Physics*, Vol. 205, 131–156, 2005.

Scaperdas, A. and S. Gilham, "Thematic Area 4: Best Practice Advice for Civil Construction and HVAC," *The QNET-CFD Network Newsletter*, Vol. 2, 28–33, 2004.

Schroeder G. and K.H. Schlünzen, F. Schimmel, "Use of (Weighted) Essentially Non-Oscillatory Advection Schemes in a Mesoscale Model, *Quarterly Journal of the Royal Meteorological Society,* Vol. 132, 1,509-1,526, 2006.

Sandia National Laboratories, "Laminar Hydraulic Analysis of a Commercial Pressurized Water Reactor Fuel Assembly," SAND2008-3938, August, 2008.

Sparrow E.M and L.F.A. Azevedo, "Vertical-Channel Natural Convection Spanning between the Fully-Developed Limit and the Single-Plate Boundary-Layer Limit," *International Journal of Heat and Mass Transfer*, Vol. 28, 1847–1857, 1985.

Sparrow E.M and A.L. Loeffler, Jr., "Longitudinal Laminar Flow between Cylinders Arranged in Regular Array," *AIChE Journal*, Vol. 5, 325–330, 1959.

Speziale, C.G., "On Non-Linear K-l and K-ε Models of Turbulence," *Journal of Fluid Mechanics*, Vol. 178, 82–89, 1987.

Speziale, C.G. and T. Ngo, "Numerical Solution of Turbulent Flow Past a Backward-Facing Step Using Non-Linear K-e Model," *International Journal of Engineering Science*, Vol. 26, 1,099-1,112, 1988.

Stern, F. and R.V Wilson, H.W. Coleman, E.G. Paterson, "Comprehensive Approach to Verification and Validation of CFD Simulations–Part 1: Methodology and Procedures," *Journal of Fluids Engineering*, Vol. 123, 793–802, 2001.
Sucec, J., *Heat Transfer*, William C. Brown Publishers, Dubuque, IA, 1985.

Thangam, S., "Analysis of Two Equation Turbulence Model for Recirculating Flows," NASA Contract Report No 187607, 1991.

TRW, "Spent Nuclear Fuel Effective Thermal Conductivity Report," prepared by TRW Environmental Safety Systems, Inc., for U.S. Department of Energy, July 11, 1996.

VDI, *Environmental Meteorology–Prognostic Microscale Wind Field Models – Evaluation for Flow around Buildings and Obstacles*, VDI Guideline 3783, Part 9, Beuth Verlag, Berlin, 2005.

Wilcox, D.C. "Turbulence Modelling for CFD," 2nd ed., DCW Industries, Inc., 1998.

Wilson, R. and J. Shao, F. Stern, "Discussion: Criticisms of the 'Correction Factor' Verification Method," *Journal of Fluids Engineering*, Vol. 126, 704–706, 2004.

Wolfshtein, M.W., "The Velocity and Temperature Distribution in a One-Dimensional Flow with Turbulence Augmentation and Pressure Gradient," *International Journal of Heat and Mass Transfer*, Vol. 12, 301–312, 1969.

WS Atkins Consultants, "Best Practices Guidelines for Marine Applications of CFD," MARNET-CFD Report, 2002.

APPENDIX A

VERIFICATION OF COMPUTATIONAL FLUID DYNAMIC CODES AND NUMERICAL SIMULATION RESULTS (Franke et al., 2007)

In the context of quality assurance of computational fluid dynamic (CFD) codes, verification deals with the relationship between the conceptual and the computerized model (Oberkampf et al., 2004). The conceptual model comprises all the equations necessary to describe the physical system, including initial and boundary conditions. The implementation of these equations into an operational computer program is called the computerized model or CFD code. Verification, therefore, is purely mathematical.

Contrary to that, validation deals with physics and is based on the comparison of the results of a numerical simulation with experimental measurements. Therefore, validation is concerned with the question of whether the conceptual models, together with the computerized model, are an appropriate representation of reality. Verification, on the other hand, is concerned solely with the question of whether the CFD code is an appropriate representation of the conceptual model. Or as Roache (1997) has formulated it succinctly, verification is used to check whether the equations are solved right and validation is used to check whether the right equations are solved.

There are two kinds of verification. One is code verification, which is used to demonstrate that the computerized model is consistent with the CFD code as stated above; that is, there are no programming errors or inconsistencies in the solution algorithm (Roy, 2005). This is normally done by the code developers. The other kind of verification is solution verification, which is the estimation of the numerical error (Roache, 1997; Oberkampf et al., 2004; Roy, 2005) or uncertainty (Stern et al., 2001) of a specific simulation result, which would be done by the code user. Solution verification is also known as numerical error estimation (Oberkampf et al., 2004).

Both kinds of verification need to quantify the discretization error that results from the fact that a system of partial differential equations (PDEs) is solved with finite discretization in space and time. The most general method for estimating the discretization error is the Richardson extrapolation (Richardson, 1910; Richardson, 1927), which is used in code verification and solution verification. Therefore, the generalized Richardson extrapolation is introduced first and, afterwards, code and solution verification are discussed in general.

A.1 Generalized Richardson Extrapolation (Franke et al., 2007)

Richardson extrapolation is a posteriori error estimator that is independent of the numerical method used to obtain the numerical solutions. It can be applied to the local flow variables, as well as to derived integral quantities. The method can be used for the spatial discretization and the temporal discretization. Here, it will be introduced for the spatial discretization. If f_{ex} is the smooth exact solution and f_k the result of a numerical solution on the mesh indexed by k, then these two can be related by a series expansion,

$$f_k = f_{ex} + g_p h_k^p + g_{p+1} h_k^{p+1} + g_{p+2} h_k^{p+2} + ... \tag{1}$$

h_k is a (linear) measure of the grid size of mesh k, p is the order of accuracy, and g are coefficients. When the solution on mesh k is in the asymptotic range, then all terms of higher order than p can be neglected and p and g do not depend on h_k (Stern et al., 2001). The only unknowns that remain on the right-hand side of Equation (1) are then f_{ex}, g_p, and p. In the most general case (which is the one encountered in solution verification), none of these is known and three equations corresponding to solutions on three different meshes are necessary to estimate f_{ex}.

If k=1 denotes the fine, k=2 the medium, and k=3 the coarse grid, two grid refinement ratios can be introduced,

$$r_{21} = \frac{h_2}{h_1}, \qquad r_{32} = \frac{h_3}{h_2} \qquad (2)$$

With these ratios, the series expansions can be written as

$$f_1 = f_{ex} + g_p h_1^p = f_{ex} + g_p h_1^p$$
$$f_2 = f_{ex} + g_p h_2^p = f_{ex} + g_p (r_{21} h_1)^p$$
$$f_3 = f_{ex} + g_p h_3^p = f_{ex} + g_p (r_{21} r_{32} h_1)^p \qquad (3)$$

The neglect of the higher-order terms in the series for the medium and coarse grid requires that these solutions are also in the asymptotic range. Another criterion for the applicability of the generalized Richardson extrapolation with solutions from three meshes is that the solution displays monotonic convergence (Stern et al., 2001). From the ratio of the solution changes, $R=(f_2-f_1)/(f_3-f_2)$, three different behaviors can be discerned.

(i) $0<R<1$: monotonic convergence
(ii) $R<0$: oscillatory convergence
(iii) $|R| >1$: divergence (4)

For divergence, no error estimate can be obtained. Oscillatory convergence generally requires the use of more solutions than three to compute an error estimate. However, the main problem with oscillatory convergence in general is that it might manifest itself as (i) or (iii) (Coleman et al., 2001). Another problem is that R may become ill-conditioned when (f_3-f_2) approaches zero. Then the maxima and minima of the local solution should be analyzed, possibly together with the ratio formed by the L_2 norms of the solution changes (Stern et al., 2001).

To calculate the solution changes, it is necessary that all solutions are available at the same positions. In case of always doubling the number of cells in each coordinate direction $(r=r_{21}=r_{32}=2)$ without moving the nodes of the coarse grid, this requirement is fulfilled. Otherwise, interpolation from the fine and medium grid on the coarse grid is necessary (Cadafalch et al., 2002). The order of the method used for interpolation must be higher than the anticipated p to not contaminate the grid convergence study (Roache, 1998). If the generalized Richardson extrapolation is applied to integral values, then no interpolation is necessary. Assuming that all solutions are available on the coarse grid and monotonic convergence according to Equation (4), the order of accuracy can be calculated from Equation (3) by solving the transcendental equation

$$p = \frac{\ln\left[\left(\dfrac{f_3 - f_2}{f_2 - f_1}\right)\right]}{\ln(r_{21})} - \frac{1}{\ln(r_{21})}\left[\ln\left(r_{32}^p - 1\right) - \ln(r_{21}^p - 1)\right] \tag{5}$$

with an iterative method. After elimination of g_p in Equation (3), an estimate of the exact solution is obtained,

$$f_{ex} = f_1 + \frac{f_1 - f_2}{r_{21}^p - 1} \tag{6}$$

The second term on the right-hand side of Equation (6) defines a correction to the fine grid solution f_1. This correction is only available at the positions of the variable on the coarse grid. To make the corrections available at every node or in every cell on the fine grid, interpolation is necessary (Roache, 1998). The error of the interpolation again has to be lower than the discretization error. The (spatial) discretization error DE_1 of the fine grid solution (i.e., the difference between the solution on the fine grid and the exact solution) follows from Equation (6):

$$DE_1 = f_1 - f_{ex} = \frac{f_2 - f_1}{r_{21}^p - 1} \tag{7}$$

For the spatial discretization errors on the medium and coarse grid, the following relations are obtained:

$$DE_2 = f_2 - f_{ex} = \frac{r_{21}^p(f_2 - f_1)}{r_{21}^p - 1} = r_{21}^p DE_1$$

$$DE_3 = f_3 - f_{ex} = \frac{r_{32}^p r_{21}^p(f_2 - f_1)}{r_{21}^p - 1} = (r_{32} r_{21})^p DE_1 \tag{8}$$

With the aid of Equation (8), it can be checked if the three solutions are in the asymptotic range. In this case, the following relation holds in which the first identity follows by definition:

$$DE_1 = \frac{DE_2}{r_{21}^p} = \frac{DE_3}{(r_{32} r_{21})^p} \tag{9}$$

The presented results are the most general form of the generalized Richardson extrapolation. They are simplified with a constant refinement ratio $r = r_{21} = r_{32}$. The order of the numerical solution then can be calculated explicitly from

$$p = \frac{\ln\left[(f_3 - f_2)/(f_2 - f)_1\right]}{\ln(r)} \tag{10}$$

A-3

The estimate from Equation (6) for the exact solution and the discretization errors from Equation (7) and Equation (8) are then:

$$f_{ex} = f_1 + \frac{f_1 - f_2}{r^p - 1} \tag{11}$$

$$DE_1 = f_1 - f_{ex} = \frac{f_2 - f_1}{r^p - 1} \tag{12}$$

$$DE_2 = f_2 - f_{ex} = \frac{r^p(f_2 - f_1)}{r^p - 1} = r^p DE_1 \tag{13}$$

$$DE_3 = f_3 - f_{ex} = \frac{r^{2p}(f_2 - f_1)}{r^p - 1} = r^{2p} DE_1$$

How the described Richardson extrapolation is used in code and in solution verification will be shown next. Here, the main prerequisites that also can be viewed as disadvantages of the method are briefly restated.

- The applicability of the method requires smooth solutions. For solutions with discontinuities or singularities, the effectiveness of the method is reduced (Roy, 2005).

- The method relies on having multiple solutions in the asymptotic range, which can be very expensive.

- The method does not work with divergent changes in the solution (see Equation (4)). Oscillatory changes in the solution might not be detected.

- The method tends to amplify other sources of numerical errors, such as round-off and incomplete iterative convergence errors. Roy (2005) states that these two errors should be at least 100 times smaller than the discretization error.

The advantages of the method are the following:

- As a postprocessing tool, it can be applied with every discretization method (finite difference, finite volume, and finite element).

- No intrusion into the code is necessary.

- The global error or estimates of this error can be calculated for every quantity.

A.2   Code Verification (Franke et al., 2007)

As already stated, code verification is used to analyze if the conceptual model is correctly implemented in the computerized model or CFD code. The correct implementation has to be demonstrated (Oberkampf et al., 2004).

If the numerical method is consistent, then the basic partial differential equations are recovered from the discrete equations in case of vanishing grid and time-step size. The rate at which the basic PDEs are approached is determined by the truncation error (e.g., if the smallest exponent of the grid size in the truncation error is 2, then the method is said to be of second-order (accuracy) in space). Halving the grid size will therefore reduce the truncation error by a factor of 4 if the solution is already in the asymptotic range, as defined above. The formal truncation error—and, thus, the formal order of the computerized model—can be found by using Taylor series expansion and subtracting the basic PDEs from the expanded discrete equations. Whether the formal order is observed in actual applications of the code is analyzed with the aid of code verification by determining the observed order of accuracy. This is the most rigorous and, therefore, recommended acceptance test for code verification (Knupp and Salari, 2003).

The observed order of accuracy is determined with the aid of Richardson extrapolation, as described above. Assuming the exact solution to the partial differential equations is known, only solutions on two meshes are required (see Equation (3)).

From these, the observed order of accuracy p can be calculated:

$$p = \frac{\ln\left[(f_2 - f_{ex})/(f_1 - f_{ex})\right]}{\ln(r_{21})} \tag{14}$$

From Equation (14), the observed order of accuracy is defined at every node in which both solutions are available. Assuming $r_{21}=2$, which is the general but not necessary choice for code verification, this requirement is fulfilled for the coarser mesh 2 without interpolation. For the verification of the code, the computation of a global discretization error suffices to calculate the observed order of accuracy. Roy (2005) describes the use of discrete L_∞ and L_2 norms, which are defined as

$$L_{\infty,k} = \max\left|f_{kn} - f_{ex,n}\right|, \qquad L_{2,K} = \left(\frac{\sum_{n=1}^{N}\left|f_{k,n} - f_{ex,n}\right|^2}{N}\right)^{1/2} \tag{15}$$

on every mesh k. Here, n is the index of the nodes or cells of the mesh and N the total number of nodes or cells. From both or one of these norms the observed order of accuracy is calculated,

$$p = \frac{\ln(L_{K+1}/L_K)}{\ln(r)} \tag{16}$$

Code verification is achieved if the observed order and the formal order coincide. There are several possible reasons for the case in which the observed and the formal order do not agree. The most important one is programming errors. Indeed, order of accuracy testing is an efficient

A-5

tool to detect these mistakes. To that end, other possible sources of disagreement between the observed and formal order of accuracy should be eliminated. These sources mainly relate to the Richardson extrapolation and are solutions that are not smooth enough, and round-off or incomplete iterative convergence errors become important. By assuring smooth solutions, as well as negligible round-off and iterative convergence errors (at least 100 times smaller than the discretization errors (see Roy, 2005)), a failure of the order of accuracy test can be safely attributed to programming errors. The method described above relies on the availability of exact solutions for the basic PDEs. Analytical solutions of the Navier-Stokes equations only exist for simple problems or are obtained after substantial simplification of the basic equations. As an alternative for the use of analytical solutions to the Navier-Stokes equations, the method of manufactured solutions (MMS) is advocated as the best choice in code verification (Roache, 2002; Oberkampf et al., 2004; Roy, 2005).

This method is based on the prescription of an analytical solution for all variables computed. These solutions do not fulfill the basic conservation equations, but they lead to additional source terms when inserted in the basic equations. Therefore, MMS does not solve the original system of equations but, rather, a modified system of equations. However, the additional terms are known and can be implemented into the code in the exact analytical form. The corresponding initial and boundary conditions also are obtained from the prescribed analytical solutions. When the original code is run with these extensions, then results of the simulation must approach the prescribed analytical solutions at a rate with the formal order of accuracy when the grid or the time step are refined. The observed order test described above, therefore, must be applied to the solutions obtained with the modified equations. As the modification (hopefully) only introduces analytical (i.e., exact terms in the code), the untouched original part of the code is tested for programming errors.

Roy (2005) summarizes code verification with MMS in the following six steps:

(1)   choice of the form of the governing equations

(2)   choice of the form of the manufactured solutions

(3)   derivation of the modified governing equations

(4)   solution of the discrete form of the modified equations on multiple meshes

(5)   evaluation of the global discretization error in Equation (15) in the numerical solutions

(6)   application of order of accuracy test to determine whether the observed order in Equation (16) matches the formal order. He also formulates the following requirements of the manufactured solutions:

   o   The analytical functions and all their derivatives should be smooth (trigonometric and exponential functions recommended). Thus, the observed order can be determined on relatively coarse meshes.

   o   The analytical functions are not allowed to lead to vanishing derivatives (also cross derivatives) in the governing equations.

   o   After insertion of the analytical functions, all terms in the original equations should be of similar orde

○ It must be certified that the analytical functions lead to realizable variable values only (e.g., the turbulent kinetic energy must be non-negative). The MMS for code verification is a powerful set of procedures to determine the correct implementation of the conceptual model in the code. It is independent of the basic discretization method (finite difference, finite volume, or finite element) and can deal with coupled sets of nonlinear partial differential equations.

○ It can also be applied to software other than CFD codes. However, MMS depends on the possibility to implement arbitrary source terms, as well as initial and boundary conditions into the code, and is therefore code intrusive. While this is certainly no problem for code developers, code users may not be able to perform code verification. Another weakness of the method is its restriction to smooth solutions. A verification example using MMS is shown in the V&V 20 standard (ASME, 2009).

A.3   Solution Verification (Numerical Error Estimation) (Franke et al., 2007)

As stated in the beginning, solution verification deals with the estimation of the numerical error or uncertainty of a given simulation result. It has been stated previously that several sources of the numerical error or uncertainty exist. This section deals only with the discretization error. Numerical errors caused by computer programming, round-off, or incomplete iterative convergence are not addressed. Rather, it is implied that these errors have been reduced to a negligible amount. The remaining numerical error can then be attributed to the finite resolution in space and time. The following methods for the estimation of this error can be applied to the space discretization and the time discretization. This discussion, however, will only describe the estimation of the spatial discretization error.

Solution verification also is performed with the aid of the generalized Richardson extrapolation. As the exact solution to the PDEs is not known, solution verification requires at least solutions on three systematically refined or coarsened meshes (i.e., the refinement or coarsening must be constant in the entire computational domain). The observed order of accuracy then can be computed from Equation (5) or Equation (10) and the discretization errors estimated from Equations (7) and (8) or Equations (12) and (13), respectively. Menter (2002) proposes to use the formal order of accuracy in the grid convergence study, thereby reducing the necessary solutions to two. However, Stern et al. (2001) state that a two-grid study only provides information about the sensitivity of the solution to the space discretization and not an error estimate. The necessity to use solutions on three meshes makes the method expensive because all three solutions must be obtained on meshes that are fine enough for the solutions to be in the asymptotic range, which has to be analyzed with Equation (9). This requirement raises the question about the minimum refinement ratio r that should be used in the grid convergence study since it determines the required number of nodes or cells. For code verification, it was stated that the ideal case is $r=2$ corresponding to a doubling of cells in each coordinate direction. This increases the number of cells from the coarse to the fine grid by a factor of 64 and therefore is very demanding of computational resources. Ferziger and Perić (2002) recommend at least an increase of 50 percent of the cells in each coordinate direction, corresponding to $r\approx3.4$. Stern et al. (2001) state that for industrial applications $r=2^{1/2}$ is an appropriate choice, and Roache (1998) shows that even $r=1.1$ is enough for simple meshes. The refinement or coarsening of the mesh is straightforward for structured meshes with hexahedral cells. The most efficient way is to start with the fine hexahedral mesh and then successively coarsen this mesh. On the other hand, for meshes with tetrahedral cells, it is easier to first generate the coarse mesh and then use refinement by subdividing every cell (Roy, 2005). On tetrahedral or unstructured meshes in general, the refinement factor r also can be defined by Roache (1998)

$$r_{21} = \left(\frac{N_1}{N_2}\right)^{1/D} \tag{17}$$

where N_k is the number of nodes or cells of the mesh and D the dimension of space. With the use of $r\neq2$, keep in mind that interpolation to the coarse grid is necessary and one must ensure that the interpolation error is smaller than the discretization error to be analyzed. The same interpolation problem arises if the correction to the fine grid solution is computed with the aid of Equations (6) or (11). Since the correction is only available on the coarse grid, it has to be transferred back to the fine grid. Besides the interpolation, another problem with the corrected solution is that it generally is no longer fulfilling the basic equations (e.g., mass conservation may not be fulfilled with the corrected fine grid solution). Therefore, the most common approach with generalized Richardson extrapolation in grid convergence studies is to calculate the relative error or an error band. This generally is done for the solution on the fine grid.

For the fine grid, the relative error is defined as

$$E_1' = \frac{f_1 - f_{ex}}{f_{ex}} \tag{18}$$

Roache (1998) has shown that this error can be approximated by

$$E_1 = \frac{1}{r_{21}^p - 1} \frac{f_2 - f_1}{f_1} = E_1' + O(h^{p+1}) \tag{19}$$

Menter (2002) suggests several practical error estimators based on Equation (19):

• Field error
$$E_{1,f} = \frac{1}{r_{21}^p - 1} \frac{|f_2 - f_1|}{range(f_1)} \tag{20}$$

• Maximum error
$$E_{1,max} = \frac{1}{r_{21}^p - 1} \frac{\max|f_2 - f_1|}{range(f_1)} \tag{21}$$

• RMS error
$$E_{1,rms} = \frac{1}{r_{21}^p - 1} \frac{rms(f_2 - f_1)}{range(f_1)} \tag{22}$$

Here, normalization of the discretization error has been performed with the range of the solution on the fine grid, defined as range (f_1)=max(f_1)−min(f_1), to exclude problems with vanishing f_1.

The field error in Equation (20) is defined at every node or cell of the coarse grid. From this error, the average error in the entire computational domain can be formed. This average error is also needed for the computation of the RMS error in Equation (22), which again gives one value for the entire computational domain, like the maximum error in Equation (21).

The magnitude of the relative error in Equation (19),

$$|E_1| = \frac{1}{r_{21}^p - 1} \left| \frac{f_2 - f_1}{f_1} \right| \tag{23}$$

which is closely related to the field error in Equation (20), defines an error band around the solution on the fine grid, i.e., $f_1 \pm |E_1|$. This definition of the error band, however, provides only 50-percent confidence that the true error falls within this error band (Roy, 2005). Therefore, the error band in Equation (23) is generally multiplied by a factor of safety F_s to increase the confidence level of the estimate

$$\left| E_{1,s} \right| = \frac{F_s}{r_{21}^p - 1} \left| \frac{f_2 - f_1}{f_1} \right| \qquad (24)$$

Roache (1994) introduced this definition and called it the grid convergence index (GCI). For F_s, he suggests two values, depending on the number of meshes used and on the relation of the observed and formal order of accuracy:

- $F_s = 1.25$ if the order of accuracy is calculated from solutions on three meshes and this observed order matches the formal one.

- $F_s = 3$ if only two meshes are used (i.e., the observed order is assumed to match the formal one, or if three meshes are used but the observed and formal order do not match).

Stern et al. (2001) derived a variable factor of safety $F_{s,c}$ based on what they called correction factor C. Their introduction of the correction factor was based on the observation that the estimate for the discretization error in Equation (7) has the correct form but that the observed order of accuracy is only poorly estimated with Equations (5) or (10) unless the results on the three meshes are in the asymptotic range. The correction factor shall remedy this problem and account for the influence of the higher-order terms neglected under the assumption that all solutions are in the asymptotic range.

Stern et al. (2001) propose two formulations for the correction factor, the simpler of which is

$$C_1 = \frac{r_{21}^p - 1}{r_{21}^q - 1} \qquad (25)$$

Here, q is an improved estimate of the order of accuracy, normally the formal order of accuracy. The correction factor therefore measures the distance of the solutions from the asymptotic range. If all the solutions used lie within the asymptotic range, then the observed order must match the formal order and $C_1 = 1$. Their factor of safety then depends on the magnitude of C (Wilson et al., 2004),

$$F_{s,c} = \begin{cases} 9.6(1 - C_1)^2 + 1.1 & |1 - C_1| < 0.125 \\ 2|1 - C_1| + 1 & |1 - C_1| \geq 0.125 \end{cases} \qquad (26)$$

For $C_1 = 1$, their factor of safety is $F_{s,c} = 1.1$, which is smaller than $F_s = 1.25$ from Roache (1994). Both factors are equal for $C_1 = (0.875, 1.125)$. Between these two intersections $F_{s,c}$ is smaller than F_s and therefore less conservative. Outside the interval defined by the intersection points $F_{s,c}$ is larger and therefore more conservative if $F_s = 3$ is not already used. The choice of the appropriate factor of safety is a matter of an ongoing discussion—especially the question of which factor would provide a 95-percent confidence level (Roy, 2005).

A-10

A.4 Application of Richardson Extrapolation to Calculate Discretization Error (Celik, 2005)

The recommended method for discretization error estimation is the Richardson extrapolation method. Since its first elegant application by its originator Richardson (1910, 1927), many authors have studied this method. Its intricacies, shortcomings, and generalization have been widely investigated. The Richardson extrapolation method is far from perfect. The local values of predicted variables may not exhibit a smooth, monotonic dependence on grid resolution, and in a time-dependent calculation, this nonsmooth response also will be a function of time and space. Nonetheless, it is currently the most robust method available for the prediction of numerical uncertainty.

The GCI method described herein is an acceptable and recommended method that has been evaluated over several hundred CFD cases. If authors choose to use it, the method will not be challenged in the paper review process. If authors choose to use another method, its adequacy will be judged in the review process. This policy is not meant to discourage further development of new methods; in fact, the ASME *Journal of Fluids Engineering* encourages the development and statistically significant evaluation of alternative methods of estimation of error and uncertainty. Rather, this policy is meant to facilitate CFD publication by providing practitioners with a method that is straightforward to apply, is fairly well justified and accepted, and will avoid possible review bottlenecks, especially when the CFD paper is an applications paper instead of one concerned with new CFD methodology.

Recommended Procedure for Estimation of Discretization Error

Step 1. Define a representative cell, mesh, or grid size h. For example, for 3-D calculations

$$h = \left[\frac{1}{N} \sum_{i=1}^{N} (\Delta V_i) \right]^{1/3} \tag{1}$$

For 2-D

$$h = \left[\frac{1}{N} \sum_{i=1}^{N} (\Delta A_i) \right]^{1/2} \tag{2}$$

where ΔV_i is the volume and ΔA_i is the area of the ith cell, and N is the total number of cells used for the computations. Equations (1) and (2) are to be used when integral quantities (e.g., heat transfer coefficient) are considered. For field variables, the local cell size can be used. Clearly, if an observed global variable is used, it is then appropriate to use also an average "global" cell size.

Step 2. Select three significantly different sets of grids and run simulations to determine the values of key variables important to the objective of the simulation study (e.g., a variable critical to the conclusions being reported). It is desirable that the grid refinement factor, *r=hcoarse/hfine*, be greater than 1.3. This value of 1.3 is based on experience and not on formal derivation. The grid refinement should, however, be done systematically; that is, the refinement itself should be structured even if the grid is unstructured. Use of geometrically similar cells is preferable.

Step 3. Let *h1<h2<h3* and *r21=h2/h1*, *r32=h3/h2*, and calculate the apparent order, *p*, of the method using the expression

$$p = \frac{1}{\ln(r_{21})}\left|\ln\left|\frac{\varepsilon_{32}}{\varepsilon_{21}}\right| + q(p)\right|, \tag{3a}$$

$$q(p) = \ln\left(\frac{r_{21}^p - s}{r_{32}^p - s}\right), \tag{3b}$$

$$s = 1 \cdot sign\left(\frac{\varepsilon_{32}}{\varepsilon_{21}}\right), \tag{3c}$$

Where $\varepsilon_{32} = \phi_3 - \phi_2$, $\varepsilon_{21} = \phi_2 - \phi_1$, ϕ_k denoting the solution on the kth grid. Note that $q(p)=0$ for *r=const.* Equation (3) can be solved using fixed-point iteration, with the initial guess equal to the first term. The absolute value in Equation (3a) is necessary to ensure extrapolation towards *h=0* (Celik and Karatekin, 1997). Negative values of $\varepsilon_{32}/\varepsilon_{21}$ <0 are an indication of oscillatory convergence. If possible, the percentage occurrence of oscillatory convergence also should be reported. Agreement of the observed apparent order with the formal order of the scheme used can be taken as a good indication that the grids are in the asymptotic range; the converse should not necessarily be taken as a sign of unsatisfactory calculations. It should be noted that if either $\varepsilon_{32} = \phi_3 - \phi_2$ or $\varepsilon_{21} = \phi_2 - \phi_1$ is "very close" to zero, the above procedure does not work.

This might be an indication of oscillatory convergence or, in rare situations, it may indicate that the "exact" solution has been attained. In such cases, if possible, calculations with additional grid refinement may be performed; if not, the results may be reported as such.

Step 4. Calculate the extrapolated values from

$$\phi_{ext}^{21} = \left(r_{21}^p\phi_1 - \phi_2\right)/\left(r_{21}^p - 1\right), \tag{4}$$

Similarly, calculate ϕ_{ext}^{32}

Step 5. Calculate and report the following error estimates, along with the apparent order *p*:

Approximate relative error:

$$e_a^{21} = \left|\frac{\phi_1 - \phi_2}{\phi_1}\right|, \tag{5}$$

Extrapolated relative error:

$$e_{ext}^{21} = \left|\frac{\phi_{ext}^{12} - \phi_1}{\phi_{ext}^{12}}\right|, \tag{6}$$

The fine grid convergence index:

$$GCI_{fine}^{21} = \frac{1.25e_a^{21}}{r_{21}^p - 1},$$

(7)

Table A-1 illustrates this calculation procedure for three selected grids. The data used is taken from Celik and Karatekin (1997), in which the turbulent 2-D flow over a backward facing step was simulated on nonuniform structured grids with total number of cells N_1, N_2, and N_3. Hence, according to Table A-1, the numerical uncertainty in the fine-grid solution for the reattachment length should be reported as 2.2 percent. (Note: This does not account for modeling errors.)

Table A-1 Sample Calculations of Discretization Error

	Φ=Dimensionless Reattachment Length (monotonic convergence)	Φ=Axial Velocity (m/s) at x/H= 8, y=0.0526 m (p<1)	Φ=Axial Velocity (m/s) at x/H=8, y=0.026 m (oscillatory convergence)
N_1, N_2, N_3	18,000, 8,000, 4,500	18,000, 4,500, 980	18,000, 4,500, 980
r_{21}	1.5	2.0	2.0
r_{32}	1.333	2.143	2.143
ϕ_1	6.063	10.7880	6.0042
ϕ_2	5.972	10.7250	5.9624
ϕ_3	5.863	10.6050	6.0909
P	1.53	0.75	1.51
ϕ_{ext}^{21}	6.1685	10.8801	6.0269
e_a^{21}	1.5%	0.6%	0.7%
e_{ext}^{21}	1.7%	0.9%	0.4%
GCI_{fine}^{21}	2.2%	1.1%	0.5%

Discretization Error Bars

When computed profiles of a certain variable are presented, it is recommended that numerical uncertainty is indicated by error bars on the profile, analogous to experimental uncertainty. It is further recommended that this be done using the GCI in conjunction with an average value of $p=p_{ave}$ as a measure of the global order of accuracy. This is illustrated in Figures A-1 and A-2. Figure A-1 (data taken from Celik and Karatekin, 1997) presents an axial velocity profile along the y-axis at an axial location of x/H=8.0 for a *turbulent* 2-D backward-facing-step flow. The three sets of grids used were 980, 4,500, and 18,000 cells, respectively. The local order of accuracy p calculated from Equation (3) ranges from 0.012 to 8.47, with a global average p_{ave} of 1.49, which is a good indication of the hybrid method applied for that calculation. Oscillatory convergence occurs at 20 percent of the 22 points. This averaged apparent order of accuracy is used to assess the GCI indices values in Equation (7) for individual grids, which is plotted in the form of error bars, as shown in Figure A-1(b). The maximum discretization uncertainty is 10 percent, which corresponds to ±0.35 m/s.

Figure A-2 (data taken from Celik and Badeau, 2003) presents an axial velocity profile along the y-axis at the station x/H=8.0 for a *laminar* 2-D backward-facing-step flow. The Reynolds number based on step height is 230. The sets of grids used were 20x20, 40x40, and 80x80, respectively. The local order of accuracy p ranges from 0.1 to 3.7, with an average value of p_{ave}=1.38. In Figure A-2, 80 percent of 22 points exhibited oscillatory convergence. Discretization error bars are shown in Figure A-2(b), along with the fine-grid solution. The maximum discretization error was about 100 percent. This high value is relative to a velocity near zero and corresponds to a maximum uncertainty in velocity of about ±0.012 m/s.

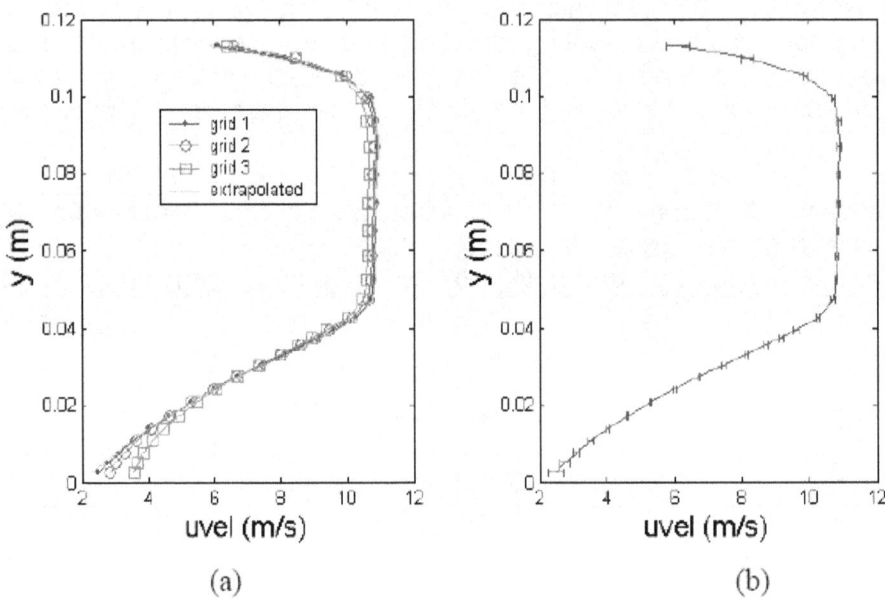

(a) (b)

Figure A-1 (a) Axial velocity profiles for a 2-D turbulent backward-facing-step flow calculation; (b) Fine-grid solution, with discretization error bars computed using Equation (7)

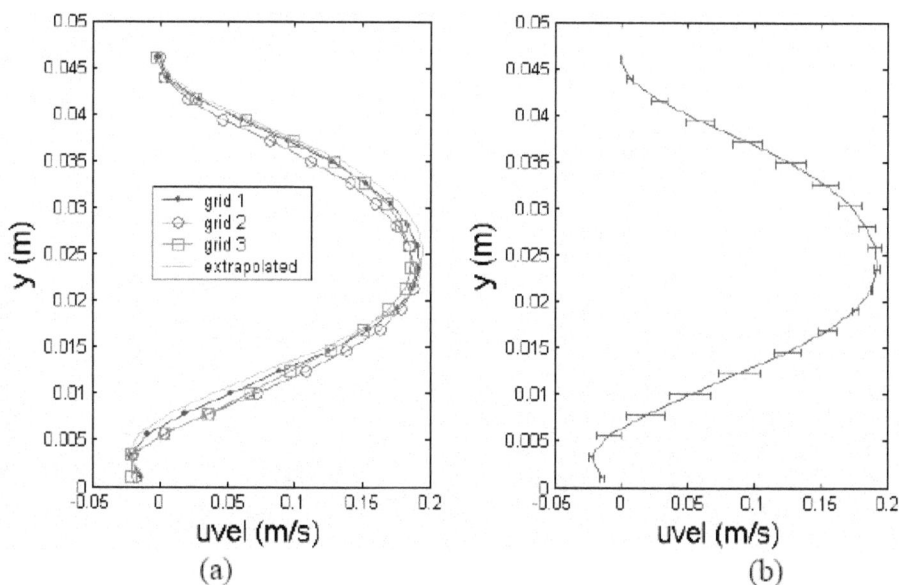

Figure A-2 (a) Axial velocity profiles for a 2-D laminar backward-facing-step flow calculation; (b) Fine-grid solution, with discretization error bars computed using Equation (7)

A-15

APPENDIX B

FLOW RESISTANCE

During fuel storage in a dry cask, helium circulates (because of buoyancy effects) from the bottom plenum through the open spaces in the fuel storage cells and exits through the top plenum and flows down through the downcomer between the basket outer walls and the canister inner shell walls. The top and bottom plenums have open spaces in the fuel basket ends to allow helium circulation. In the case of boiling-water reactor (BWR) fuel storage, a channel enveloping the fuel bundle divides the flow in two parallel paths. One flow path is through the in-channel or rodded region of the channel box and the other flow path is in the square annulus between the channel and the cell walls. In the pressurized-water reactor fuel canister, there is no channel box. Only the cell box surrounds the fuel bundle.

Different modeling approaches represent the fuel canister. One approach is to represent explicitly each fuel rod in the assembly and all the walls surrounding the rods. Another is to use porous media to model the flow within the storage canister. The extent of porous media use depends on the user's geometrical simplification of the real configuration. Some users prefer to use three-dimensional (3-D) porous media to represent only the rods in each assembly, explicitly modeling each wall in the assembly. Others prefer to use 3-D porous media to model all the assemblies, including walls between assemblies. A third option is to use an axisymmetric model to replace the entire fuel basket and the stored spent nuclear fuel by a porous cylinder. The first option is the most accurate representation, without introducing any geometrical simplification, but it is CPU (central processing unit) extensive and time consuming. The third option is quick, but may not be an accurate way to represent a dry cask and obtain best estimate predictions. The second option is a compromise and a balanced approach between the first and the third option. This option uses porous media to represent the fuel rods in each assembly, instead of representing all the assemblies in one homogenized volume.

To use porous media in computational fluid dynamics (CFD), frictional and inertial resistance parameters are needed to replace the elements omitted and simplified from the original real geometry. To obtain these porous media flow resistance parameters, the user can either use a 3-D CFD analysis or experimental data for pressure drop from Sandia National Laboratories (2008). In this appendix, CFD was used to obtain these resistance parameters for the geometry described as the second approach, as shown in Figures B-1 through B-3. In this example a BWR 10x10 assembly is used. As shown in Figures B-1 through B-3, fuel rods, water rods, grid spacers, and flow passages are explicitly included in the CFD model. The case should reflect and model flow losses in the expected operating conditions (pressure and average gas temperature) when it is inside the dry storage cask. In the present analysis, a pressure of 7 atm (standard atmosphere) and a temperature of 505 Kelvin (K) were used. In this example, two separate simplifications were used in the BWR fuel assembly dry cask cell. In the first simplification, the channel-to-cell gap was explicitly modeled as shown in Figure B-4(a), and the porous media occupied the volume enclosed by the channel box. In the second simplification, the gap between the channel and the storage cell walls was not explicitly modeled and the porous media was used to model the entire volume enclosed by the storage cell walls as shown in Figure B-4(b). Porous media parameters for both simplifications were calculated and tabulated in Table B-1.

In the prepared CFD model, the fuel storage cell length between the bottom and top plenums is replaced by porous media. To characterize the flow resistance of fuel assemblies inside the fuel channel for BWR fuel, a 3-D model of GE 10x10 fuel assemblies is constructed using the FLUENT CFD program as shown in Figures B-1 through B-3. This model represents explicitly the fuel rods, water rods, and grid spacers. Two approaches were used to calculate the flow resistance parameters: the pressure drop method and shear stress method. Both methods are applied for sections without flow area changes (i.e., no contractions or expansions). Both approaches are related and should lead to the same values. The approach to obtain the flow resistance values used in both the 3-D and the axisymmetric model is described below.

FLUENT porous media flow resistance model is defined by FLUENT (2006):

$$\frac{\Delta P}{L} = D\mu V + C\left(\frac{1}{2}\rho V^2\right)$$ (1)

Where

ΔP is the porous media pressure drop
V is the superficial fluid velocity
L is the length of porous media
μ is the fluid viscosity
ρ is the fluid density
D is the viscous resistance parameter
C is the inertial resistance parameter

In dry cask application, the C factor is not as dominant as the D factor because of the low velocity that exists inside the canister. As such, the entire pressure drop was assumed to be entirely caused by frictional losses. As a verification, the inertial coefficient (C) can be computed from correlations using area contractions and expansion (Idelchik, 1993) in the assembly to show that the second term in Equation (1) is negligible. Additionally, it would be conservative to neglect C because predicted peak cladding temperatures will be slightly higher. By definition the frictional pressure drop is:

$$\frac{\Delta P}{L} = \frac{f}{D_h}\frac{1}{2}\rho V^2$$ (2)

Where D_h is the hydraulic diameter.

Knowing that:

$$\text{Re} = \frac{\rho V D_h}{\mu}$$

We get:

$$\frac{\Delta P}{L} = \frac{f\,\text{Re}}{2}\frac{\mu}{D_h^2}V$$

Usually the friction factor in the laminar regime, as shown in a Moody diagram, will have the following form:

$$f = \frac{A}{\text{Re}} \tag{3}$$

As an illustration, the frictional coefficient because of the pressure drop for a *laminar flow pipe* has been experimentally determined to correspond to the following expression:

$$f = \frac{64}{\text{Re}} . \quad \text{Thus:} \qquad \frac{\Delta P}{L} = \frac{32\mu}{D^2_h} V$$

For an *array of solid rods*, as is the case of BWR nuclear spent fuel assembly, the value of factor "A" can be determined from available literature (Sparrow and Loeffler, 1959). "A" has been found to have a value around 100, depending on the pitch to diameter ratio (p/d) and the porosity of the array. Using Equation (1) and neglecting the inertial term caused by the low velocities existing inside the canister, it can be concluded that the dominant contributor to pressure drop is viscous effects. The pressure drop through the rod array can be simplified to:

$$\frac{\Delta P}{L} = D\mu V \tag{4}$$

Then

$$D = \frac{A}{2D_h^2} \tag{5}$$

For laminar flow inside a pipe A=64 and the input frictional resistance in FLUENT should be:

$$D = \frac{32}{D_h^2}$$

Also, by definition:

$$f = \frac{4\tau_w}{\frac{1}{2}\rho V^2} \tag{6}$$

Where τ_w is the wall shear stress.

The porous media frictional flow resistance values for D were calculated by using both pressure drop and shear stress. Both methods should lead to similar results. The viscous resistance parameter D using the shear stress output data from the CFD analysis is obtained using the combination of Equations (2), (4), and (6). The following expression is obtained:

$$D = \frac{4\tau_w}{\mu V D_h}$$

If the pressure loss data were used, the expression for D is obtained from Equation (4) as follows:

$$D = \frac{\Delta P}{L \mu V}$$

From the CFD calculations, the wall shear stresses and pressure drop values should be obtained separately for bare fuel rods and fuel rods plus grid straps. The calculated frictional porous media flow resistance parameters are provided in Table B-1.

Table B-1 Flow Resistance Factors to Use with Porous Media in CFD

Region	First Simplification (1/m²)	2nd Simplification (1/m²)
Active region	1.014E6	1.442E6

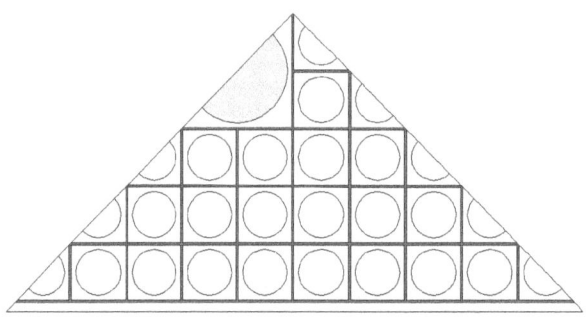

Figure B-1 One-fourth cross section of BWR 10X10 fuel assembly at the space grid level

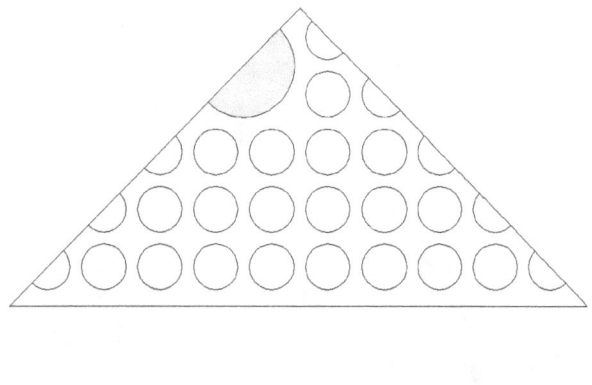

Figure B-2 One-fourth cross section of BWR 10X10 fuel assembly between space grids

Figure B-3 Longitudinal view of BWR 10X10 fuel assembly showing fuel rod and grid spacers regions

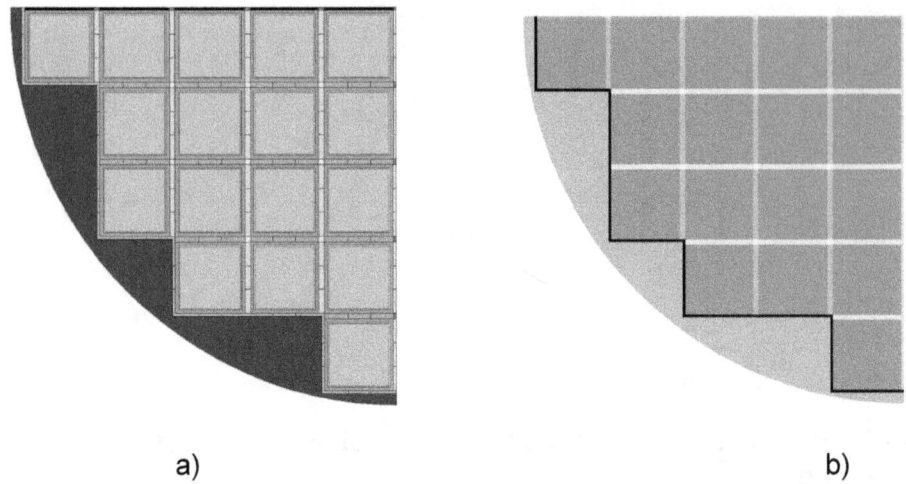

Figure B-4 One-fourth cross section of a 68 fuel assembly basket using porous media modeling a) inside the channel box; b) inside the cell box

APPENDIX C

PUBLIC COMMENTS RECEIVED AND THEIR DISPOSITION

The purpose of this appendix is to list all the public comments received on draft NUREG-2152 "Computational Fluid Dynamics Best Practice Guidelines for Dry Cask Applications." The NRC issued draft NUREG-2152 (ML12286A301) for public comment on October 23, 2012 for a 60 day period and received comments from the following two sources:

- Pacific Northwest National Laboratory (PNL), Harold E. Adlkins, Jr. Senior Research Engineer, and James A. Fort, 902 Battelle Boulevard, Richland, WA (509) 372-6629 (ML12347A191)
- Idaho National Laboratory (INL), Sandra M. Birk, Ph.D., Department Manager, Nuclear Materials Disposition & Engineering, P.O. Box 1625, Idaho Falls, ID (208) 526-1866 (ML12347A192)

The staff's resolution and any associated changes to the standard review plan are listed for each comment.

Comment:	**PNL 1**
NUREG-2152: Location	Abstract
Public Comment:	In the Abstract, it is stated that, "This report provides an independent verification and validation of the modeling approaches used to represent the heat transfer and fluid flow in a dry cask to reduce modeling uncertainties." It may be more accurate to replace "independent verification and validation" with "example application where validation".
Resolution:	The staff agrees to partly change the Abstract.
Change:	The Abstract is changed to state: This report provides validation of the modeling approaches used to represent the heat transfer and fluid flow in a dry cask to reduce modeling uncertainties
Comment:	**PNL 2**
NUREG-2152: Location	Chapter 1
Public Comment:	In the Introduction, reference to 'CFD users in the maritime industry' seems out of place for the intended user of this document.
Resolution:	The statement implies that the physics and equations are common to all CFD applications. In other words, this application deals with the same issues involving verification and validation.
Change:	No change
Comment:	**PNL 3**
NUREG-2152: Location	Chapter 1
Public Comment:	Also in the Introduction, the list of areas of general CFD that overlap with those associated with dry cask applications include" multiphase phenomena, chemical species interaction, and combustion". These specifically would not likely be involved with dry cask applications and should be removed from the list.
Resolution:	For extended storage one could be dealing with multiphase flow, involving humidity deposition on the canister shell.
Changes:	No change

Comment:	**PNL 4**
NUREG-2152: Location	Chapter 1
Public Comment:	In the same paragraph as the previous comment, insert the word 'careful' in, "because of the complexity of modern commercial CFD packages, [careful] input preparation and solution of equation modeling are essential to avoiding errors".
Resolution:	The staff agrees to change the paragraph and to modify the statement for more clarity.
Changes:	Chapter 1 is changed to state: careful input preparation and solution of model equations are essential to avoiding errors.
Comment:	**PNL 5**
NUREG-2152: Location	Chapter 3
Public Comment:	In Section 3.1.5, on page 18, the four bullets listing priorities in mesh types would be more clear if shown as a numbered list.
Resolution:	Bullets are used in all sections so the proposed changed would result in inconsistent format for entire document.
Changes:	No change
Comment:	**PNL 6**
NUREG-2152: Location	Chapter 3
Public Comment:	In Section 3.2.2, on page 21, insert the word 'features' in the sentence, "If time steps are too large, the simulation might fail to capture important flow [features] and mimic unphysical steady behavior."
Resolution:	The staff agrees with the change.
Changes:	Chapter 3 is changed to state: If time steps are too large, the simulation might fail to capture important flow features and mimic unphysical steady behavior.
Comment:	**PNL 7**
NUREG-2152: Location	Chapter 5
Public Comment:	Several places in the document refer to the phenomena identification and ranking (PIRT) process, but there isn't a detailed description of what this

is. Suggest adding a reference for readers not familiar with the PIRT process.

Resolution:	The staff added reference to NUREG/CR-6978
Changes:	Chapter 5 is changed to state: Numerical errors should be monitored for a limited number of representative target variables defined during the phenomena Identification and ranking table (PIRT) process (NUREG/CR-6978, 2008)

Comment: **PNL 8**

NUREG-2152: Location	Chapter 3
Public Comment:	There are two "Spatial Discretization Errors" sections (3.21 and 5.4) and likewise two "Time Discretization Errors" sections (3.22 and 5.5). The second of these deal more with error estimation and the former is more general describing tradeoffs. Suggest removing 'Errors' from the titles of Sections 3.21 and 3.22.
Resolution:	The staff agrees with the change but the titles of Sections 3.2.1 and 3.2.2 are modified to better agree with the content of these sections.
Changes:	Chapter 3 is changed to state: 3.2.1 Spatial Discretization Schemes and 3.2.2 Time Discretization Schemes

Comment: **PNL 9**

NUREG-2152: Location	Chapter 7
Public Comment:	In the Application section 7.1.1, correct table reference to Table 7-2.
Resolution:	The staff intended to point the reader to Table 2 of McKinnon et al., 1992. In reality Table S-2 should be referenced.
Changes:	Chapter 7 is changed to state: Table S-2 (McKinnon et al., 1992)

Comment: **PNL 10**

NUREG-2152: Location	Chapter 7
Public Comment:	In the Application Section 7.1.7, reference is made to Succec (1985). This reference is not included in the Reference list.
Resolution:	The author's name is misspelled in the Reference list.
Changes:	Chapter 7 is changed to state: The most complete set of data are listed in Sucec (1985)

Comment:	**PNL 11**
NUREG-2152: Location	Chapter 7
Public Comment:	The final bullet in the Application conclusions, section 7.2, is incompletely stated.
Resolution:	The staff agrees with the comment. Bullet is modified to provide a clear statement.
Changes:	Chapter 7 is changed to state: For higher elevation both ambient pressure and air mass flow rate decrease. Subsequently, PCT will increase.
Comment:	**PNL 12**
NUREG-2152: Location	References
Public Comment:	A reference to Appendix B could not be found within the body of the report.
Resolution:	The staff agrees with the comment
Changes:	Chapter 8 is changed to state: Were the correct inertial and frictional loss coefficients used? (See Appendix B for additional information on how to evaluate the inertial and frictional loss coefficient factors)
Comment:	**INL 1**
NUREG-2152: Location	Chapter 6
Public Comment:	On page 62 at the third bullet under Guideline on Weakness of the Standard k-e model, please add the following at the end of sentence.

Realizable k- ε turbulence model and shear stress transport (SST) k-w model are also recommended to use as part of the sensitivity study. Realizable k-ε model intended to address deficiencies of traditional k-e model by adopting a new eddy-viscosity formula and a new model equation for dissipation based on the dynamic equation for the mean-square vorticity fluctuation.

SST k-w model incorporates a damped cross-diffusion derivative in the w equation and it includes a blending function to ensure that equations behave appropriately in both the near-wall and far-field zones from the wall sublayer. |
| Resolution: | Isotropic (linear) turbulent models will not predict the complex strain fields adequately. Anisotropic models like RSM and non-linear k- ε model will be a good predictor, as indicated in draft NUREG-2152 |

Changes:	No change
Comment:	**INL 2**
NUREG-2152: Location	Chapter 3
Public Comment:	The following section is recommended to add 3.1.5

3.1.6 Grid sensitivity and convergence

The objective of this CFD calculation is to validate the experiment. It is therefore necessary to achieve the best prediction with less numerical error either in a spatial or temporal dimension. The Richardson extrapolation method is a well known method to validate the CFD result and give asymptotic value at zero spacing (Roche 1998).

In order to prevent numerical diffusion and assure a low courant number as shown in Equation 1, Richardson extrapolation method is recommended.

Courant Number = $V\Delta t/\Delta x$ (1)

where

V   = Fluid velocity (m/s)

Δt = Time step (sec)

Δx = Mesh length (m).

With grid triplet (set of fine mesh, normal mesh, and coarse mesh) solutions, the unknown variables A, B, and p ought to be obtained from simple algebra as shown in Figure 1 where p is the order of convergence in the Richardson extrapolation curve and A is the predicted value of CFD simulation result at asymptotic zero spacing grid. With constant p value, two solutions from normal grid and coarse grid allow the A value (Asymptotic front heat speed) to be calculated as:

$$p = \frac{\ln\left(\dfrac{y_3 - y_2}{y_2 - y_1}\right)}{\ln\dfrac{x_2}{x_1}} \qquad (2)$$

$$y = A + Bx^p, p, (x_2, y_2), (x_3, y_3) \text{ are known} \qquad (3)$$

$$\ln(y - A) = \ln(B) + p\ln(x) \qquad (4)$$

$$\ln\left(\frac{y_2 - A}{y_3 - A}\right) = \ln\left(\frac{x_2}{x_3}\right)^p \quad \text{3-4)}$$

$$A = \frac{0.5^p xy_3 - y_2}{0.5^p - 1}, \frac{x_2}{x_3} = 0.5 \qquad (5)$$

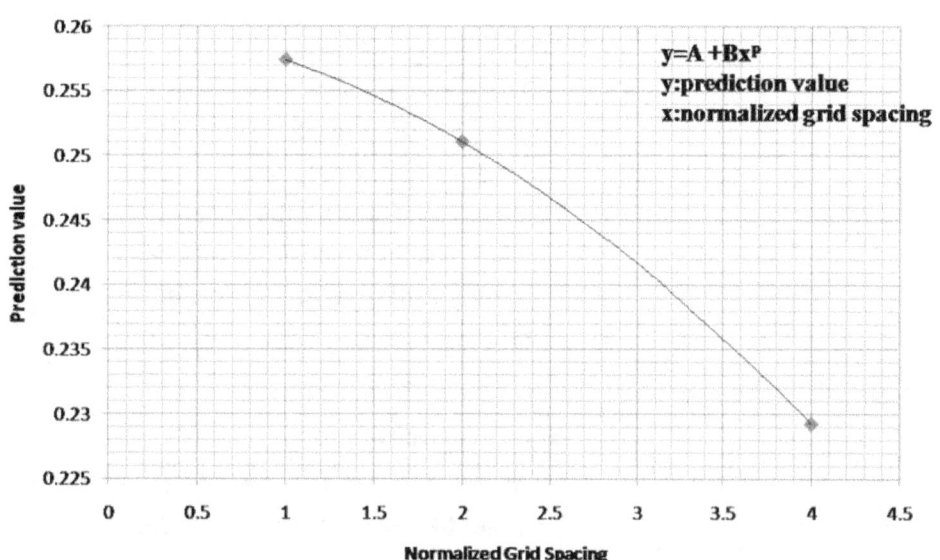

Figure 1. Correlation of grid spacing and prediction value in grid conversion.

where p = order of convergence, y_1 = variable in coarse grid, y_2 = variable in normal grid, y_3 = variable in coarse grid, and x = normalized grid spacing. If p is greater than unity, it converges while p values less than unity make the solution diverge.

The uncertainty of CFD prediction is also performed based on the Grid Convergence Index (GCI) method (Roach 1998) to make certain that solutions are checked within the asymptotic range of convergence, which should be close to unity. An example of these calculations is shown in Oh and Kim (2010).

Reference
Roache, P. J., Verification and Validation in Computational Science and Engineering, Hermosa Publishers, Albuquerque, New Mexico, 1998.
Oh, C.H. and Kim, E.S., "validation of CFD code for density-gradient driven air ingress stratified flow," Proceedings of the 18th International Conference on Nuclear Engineering (ICONE 18), Xian, China, May 17-21, 2010.

Resolution:	This comment is treated in Chapter 5 and in more detail in Appendix A
Changes:	No change
Comment:	**INL 3**
NUREG-2152: Location	Chapter 5
Public Comment:	In section 5 of V&V of the calculation and numerical method, the following subsection is recommended. The CFD specification and models recommended are listed as follows:

Solver:
- Solver: pressure based
- Formulation: implicit
- Space: 3-D double precision
- Time: steady or unsteady
- Velocity formulation: absolute
- Gradient I option: green-gauss cell based
- Unsteady formulation: 2nd-order implicit
- Pressure-velocity coupling: PISO.

Discretization:
- Pressure: PRESTO!
- Momentum: 2_{nd}-order upwind
- Turbulent kinetic energy: 2_{nd}-order upwind
- Turbulent dissipation rate: 2_{nd}-order upwind
- Species: 2_{nd}-order upwind
- Energy: 2_{nd}-order upwind.

Viscous Model:
- Turbulence model: realizable k-e or RSM
- Wall function: standard wall function.

Energy equation.

Species transport model:
- Mixture material: Mixture-template

Radiation heat transfer model

Resolution:	The suggested guidelines are not adequate for all CFD problems. The guideline's intent is to provide flexibility to the user in carefully selecting the model (including sensitivity calculations)
Changes:	No change
Comment:	**INL 4**
NUREG-2152: Location	Chapter 7
Public Comment:	Subsection 7.3 Regulatory Requirements is recommended to add.

7.3 Regulatory Thermal Requirements

This section identifies Title 10 of the Code of Federal Regulations (CFR) Part 72 relevant to the review areas, in particular, for material temperature limits. According to the ISG-11 (cladding consideration for the transportation and storage of spent fuel, interim guidance-11, revision 3), NUREG-1567, and 10CFR72,128(a), the temperature limits of the PWR fuel cladding should be within normal operating condition and accident condition for 20 years dry storage for ISFSI or MRS design and environmental condition at 400°C and 570°C, respectively.

Resolution:	This is already covered in the abstract. There is no need to repeat it again in Chapter 7
Changes:	No change
Comment:	**INL 5**
NUREG-2152: Location	Chapter 7
Public Comment:	This section describes CFD calculations on VSC-17. We recommend the following:
	Section 7.1.3 CFD modeling needs to be created and the current subsections from 7.1.3 to 7.1.10 need to be 7.1.3.1, 7.1.3.2 and so on.
	Under 7.1.3, FLUENT specific models that were used in these calculations need to be included here in terms of solver, discretization, viscous model, and etc.
Resolution:	All comments related to specific models used in the VSC-17 validation are covered in Section 7.1.3 through 7.1.10.
Changes:	No change
Comment:	**INL 6**
NUREG-2152: Location	Chapter 7
Public Comment:	Under 7.2 conclusions, the following sentence needs to be added.
	The CFD predictions on the liner and MSB wall axial temperatures are not accurate and more turbulence models such as RSM and realizable k-ε model are recommended.
Resolution:	This information is already included in the first 7 bullets of Section 7.2 in more detail.
Changes:	No change
Comment:	**INL 7**
NUREG-2152: Location	Chapter 8

Public Comment:	Are the ratios of adjacent volumes less than 2? This volume ratio of 2 is too big. The volume should be gradually increased from the fine mesh and the recommended ratio is 1.2. If the ratio of 2 is used, there is no accuracy in its computation.
Resolution:	The recommended ratio of 1.2 is a line stretching ratio not a volume expansion ratio. This will lead a volume expansion ratio of less than (or close to) 2
Changes:	No change
Comment:	**INL 8**
NUREG-2152: Location	Chapter 8
Public Comment	Are the aspect ratios below the values given in the solver manual (typically, 10-50)? This aspect ratio of 10-50 is too big. The recommended aspect ratio should be less than 20.
Resolution:	A value of 20 lies inside the recommended range.
Changes:	No change
Comment:	**INL 9**
NUREG-2152: Location	Chapter 8
Public Comment:	Is one dealing with laminar, transitional, or turbulent flow? It is turbulent flow. But the flow inside the MSB is laminar.
Resolution:	Sometimes it is not necessarily true. The staff's position is to leave it to the user to determine an adequate flow regime.
Changes:	No change
Comment:	**INL 10**
NUREG-2152: Location	Chapter 8
Public Comment:	Does the choice of turbulence level adequately represent the study objective? The choice of the turbulence model depends on the application. RSM, realizable k- ϵ, or SST k-w turbulence models are recommended.
Resolution:	It is not necessarily true. The choice of turbulent model depends on the type of flow
Changes:	No change

Comment: **INL 11**

NUREG-2152:   Chapter 8
Location

Public Comment: Is the selected wall turbulence model appropriate with the flow features?
When the wall function is used, y-plus values are very important to model
the sublayer of the wall for the turbulence model.

Resolution: The staff agrees with the comment. However, this is not the intent of the
generic question. Guidelines on turbulence models are provided in
Section 6.6.

Changes: No change

Comment: **INL 12**

NUREG-2152:   Chapter 8
Location

Public Comment: Was the first-order upwind spatial discretization avoided unless
necessary (to avoid numerical diffusion)?
The second-order upwind is recommended and also grid sensitivity study
has to be performed along with grid convergence index calculations to
avoid the numerical diffusion.

Resolution: The intent of this bullet is of generic nature. Detailed guidelines are
provided in Sections 3.2, 5.4, and 5.5.

Changes: No change

NRC FORM 335
(12-2010)
NRCMD 3.7

U.S. NUCLEAR REGULATORY COMMISSION

BIBLIOGRAPHIC DATA SHEET

(See instructions on the reverse)

1. REPORT NUMBER (Assigned by NRC, Add Vol., Supp., Rev., and Addendum Numbers, if any.)
NUREG-2152 FINAL

2. TITLE AND SUBTITLE	3. DATE REPORT PUBLISHED	
Computational Fluid Dynamics Best Practice Guidelines for Dry Cask Applications	MONTH	YEAR
	March	2013
	4. FIN OR GRANT NUMBER	

5. AUTHOR(S)	6. TYPE OF REPORT
Ghani Zigh and Jorge Solis	Technical
	7. PERIOD COVERED (Inclusive Dates)

8. PERFORMING ORGANIZATION - NAME AND ADDRESS (If NRC, provide Division, Office or Region, U. S. Nuclear Regulatory Commission, and mailing address; if contractor, provide name and mailing address.)

Division of Spent Fuel Storage and Transportation
Office of Nuclear Material Safety and Safeguards
U.S. Nuclear Regulatory Commission
Washington, D.C. 20555-0001

9. SPONSORING ORGANIZATION - NAME AND ADDRESS (If NRC, type "Same as above", if contractor, provide NRC Division, Office or Region, U. S. Nuclear Regulatory Commission, and mailing address.)

10. SUPPLEMENTARY NOTES

11. ABSTRACT (200 words or less)

Dry storage cask designs for spent nuclear fuel are submitted to the U.S. Nuclear Regulatory Commission (NRC) for certification under Title 10 of the Code of Federal Regulations (10 CFR) Part 72, "Licensing Requirements for the Independent Storage of Spent Nuclear Fuel, High-Level Radioactive Waste, and Reactor-Related Greater Than Class C Waste." The NRC staff technical review of these designs is performed in accordance with 10 CFR Part 72 and the "Standard Review Plan (SRP) for Spent Fuel Dry Storage Systems at a General License Facility" (NUREG 1536, 2010). To ensure that the cask and fuel material temperatures of the dry cask storage system will remain within the allowable limits or criteria for normal, off normal, and accident conditions, a thermal review is performed as part of the application's technical review. In cooperation with the Division of Spent Fuel Storage and Transportation of the Office of Nuclear Material Safety and Safeguards, the Office of Nuclear Regulatory Research developed this guide to provide practical advice for reviewing CFD methods used in vendor applications and for achieving high quality CFD simulations of a dry cask. To assist in the analysis, the report includes procedures, analysis methods, and acceptable assumptions.

12. KEY WORDS/DESCRIPTORS (List words or phrases that will assist researchers in locating the report.)	13. AVAILABILITY STATEMENT
CFD Best Practice Guidelines	unlimited
Standard Review Plan	14. SECURITY CLASSIFICATION
SRP	*(This Page)*
CFD	unclassified
	(This Report)
	unclassified
	15. NUMBER OF PAGES
	16. PRICE

NRC FORM 335 (12-2010)

Printed
on recycled
paper

Federal Recycling Program

UNITED STATES
NUCLEAR REGULATORY COMMISSION
WASHINGTON, DC 20555-0001

OFFICIAL BUSINESS

NUREG-2152
Final

Computational Fluid Dynamics Best Practice Guidelines
for Dry Cask Applications

March 2013